IN SEARCH OF MECHANISMS

IN SEARCH OF MECHANISMS

Discoveries across the Life Sciences

CARL F. CRAVER and

LINDLEY DARDEN

The University of Chicago Press

Chicago and London

Carl F. Craver is associate professor in the Philosophy-
Neuroscience-Psychology Program at Washington University
in St. Louis. **Lindley Darden** is professor of philosophy at the
University of Maryland.

The University of Chicago Press, Chicago 60637
The University of Chicago Press, Ltd., London
© 2013 by The University of Chicago
All rights reserved. Published 2013.
Printed in the United States of America

22 21 20 19 18 17 16 15 14 13   1 2 3 4 5

ISBN-13: 978-0-226-03965-7 (cloth)
ISBN-13: 978-0-226-03979-4 (paper)
ISBN-13: 978-0-226-03982-4 (e-book)

Library of Congress Cataloging-in-Publication Data

Craver, Carl F., author.
    In search of mechanisms : discoveries across the life
sciences / Carl F. Craver and Lindley Darden.
        pages cm
    Includes bibliographical references and index.
    ISBN 978-0-226-03965-7 (cloth : alkaline paper) —
ISBN 978-0-226-03979-4 (paperback : alkaline paper) —
ISBN 978-0-226-03982-4 (e-book) 1. Mechanism
(Philosophy) 2. Biology—Philosophy—History.
I. Darden, Lindley, author. II. Title.
    QH331.C898 2013
    146'.6—dc23

                                    2013003717

Dedicated
to
Anna
and
Lindley Joy,
Amanda,
Amelia, and
Elliott

SUMMARY OF CONTENTS

Contents

ILLUSTRATIONS

PREFACE

Science is an engine of discovery. From the farthest reaches of space to the most fundamental units of matter, the crowning aim of science is to open the black box of nature and to show how it works. In this book, we use examples drawn from across the life sciences to illustrate how biologists open black boxes to reveal the hidden mechanisms that produce, underlie, or maintain the phenomena characteristic of living things.

We have three primary objectives. First, we offer a systematic, *descriptive account* of historical and contemporary episodes illustrating biologists' search for mechanisms. We discuss the questions that figure in the search for mechanisms, and we characterize the experimental, observational, and conceptual considerations used to answer them. We also characterize the dimensions of success along which scientists measure, or ought to measure, their progress. Second, we offer an *instructive framework* to provide advice for learning how to reason about mechanisms. We use examples from the history of biology to highlight questions, constraints, and strategies that scientists use or might use as they construct, evaluate, and revise their descriptions of mechanisms. We characterize the strategies abstractly so that they might be suitable for application in altogether different discovery problems. Third, we offer a vision for the *integration of biology*. The search for mechanisms serves as one of the central integrative ideals for the biological sciences. Biology is in large part a search for mechanisms, and the search for mechanisms typically requires interfield collaboration. We show how the diverse fields of biology, from evolutionary biology to protein chemistry, from anatomy to ecology, from neuroscience to oncology, integrate their differing perspectives when they contribute jointly to an understanding of a complex mechanism.

We write this book primarily for those who like to discover things for themselves and for those who are teaching people to discover things for themselves. We provide a clear picture of what the search for mechanisms is and of how the search for mechanisms is driven by empirical investigation. We have also compiled a catalogue of resources to consider during a discovery episode. We hope that our colleagues in the history and philosophy of science will see in this work a novel vision of the nature of discovery in biology, a much neglected topic. But our primary goal is to write a practically useful book for those engaged in discovery: one that brings some order and reason to the messy process of searching for mechanisms and that offers some inspiration and caution from the history of science.

Most fundamentally, then, this book is for the curious. It is for people who enjoy figuring out for themselves how things work. Most people, we are told, have only a very shallow understanding of even rudimentary facts about biology, and even about how the mundane items in their lives (such as televisions and toasters) work. The curious, in contrast, are unsettled when they discover such gaps in their knowledge, and they set out to fill them. Most people, we are told, are utterly unaware of the gaps in their knowledge and confidently report to know quite well certain rudimentary facts about biology (and about televisions and toasters) when they do not know them at all. The curious are unsettled when they discover such lapses in their judgment, and they develop ways of being vigilant against them.

We describe the discovery of mechanisms and a set of general questions, constraints, and strategies exhibited in discovery episodes across the life sciences. We show these questions, constraints, and strategies at work in classic exemplars of research from fields across biology and through the centuries. To the extent that scientific reasoning about mechanisms represents a most exacting and refined exemplar of how properly to learn for one's self how things work, the various historical examples we have chosen function as exemplars of mechanism discovery, specimens that offer clues about how to discover mechanisms. Taken collectively, the examples establish the tremendous reach of the mechanistic worldview in contemporary biology and the centrality of the search for mechanisms to much of what biologists do.

In order to facilitate communication with an audience beyond our disciplinary homes, we wrote this book with a few explicit rules. We must say a bit to justify our decisions.

The first rule is: Stay positive. Our goal is to provide the framework for a productive and useful philosophy of discovery grounded in our previous work on mechanisms and mechanistic explanation. We measure progress in terms of the clarity with which we describe the diverse aspects of mechanism discovery, the diversity and utility of the discovery strategies we emphasize, and the suggestive pull of our approach for those who would work with us to take this admittedly preliminary project further. This is not a book about philosophical disagreements. It is a book about scientific discovery.

The second rule: Minimize disciplinary self-reference. For example, we avoid when possible the proprietary jargon of philosophy and favor plain descriptive terms. Likewise, we avoid the studious contextualization, footnoting, and qualification characteristic of academic writing in order to focus on the reasoning strategies on display in our exemplars of mechanism discovery.

Most importantly, we have removed from our text almost all references to the primary and secondary literature on which we rely. Many of the ideas in this book came directly out of our own studies in the history and philosophy of biology. Some came from our own scientific research experience. Some came from reading such sources as Francis Bacon, René Descartes, Robert Boyle, Claude Bernard, Ramon y Cajal, Francis Crick, and Joshua Lederberg. Others came from our contemporary colleagues in history and philosophy who emphasize the importance of mechanisms, most notably: Adele Abrahamsen, Bill Bechtel, Jim Bogen, Stuart Glennan, Peter Machamer, Paul Thagard, Robert Richardson, and Bill Wimsatt. Our goal is to collect many of the insights from these diverse literatures into a synthetic whole.

One perhaps controversial consequence of this choice is that we have as a rule avoided giving explicit credit within the main chapters to the colleagues from whose work we have drawn. In part to compensate for this choice, we include at the end of each chapter a section to acknowledge the sources for our ideas and to suggest further reading. It will be obvious to anyone who knows this literature how much we owe to the others who work on the nature of scientific discovery and the role of mechanisms in science. We apologize if there are pieces of the now-burgeoning literature on mechanisms that we have failed adequately to acknowledge. These bibliographic sections also contain references to the primary and secondary sources for the examples we discuss.

A final rule of this book is: Use diverse examples. Among those judged to be worthy of extended discussion, some are mandatory classics of discovery in the history of biology: Darwinian evolution by natural selection, Mendelian heredity, the evolutionary synthesis, and the discovery of the double helix structure of DNA and its role in the mechanism of protein synthesis. However, we also wanted to showcase some historical exemplars that have not received sustained philosophical discussion, such as: Harvey's theory of the circulation of the blood, Hodgkin and Huxley's model of the action potential, Loewi's discovery of neurotransmitter mechanisms, and Karl Deisseroth's development of optogenetics. These examples reflect our emphasis on the much-neglected *physiological* wing of the life sciences: a wing of biology dedicated to understanding the mechanisms by which living systems work. Yet we have also chosen examples from other areas of biology as well, such as ecology, embryology, epidemiology, and bacterial genetics, to show the extensive reach of the mechanistic perspective in biology.

If science is the engine of discovery, perhaps we can decompose it into its component parts to reveal the inner machinery by which discoveries are made. It is our hope that by making these aspects of mechanism discovery explicit, we might provide resources for both the curious and for those charged with the responsibility of guiding the curious into productive paths of future research.

ACKNOWLEDGMENTS

Lindley Darden thanks Peter Machamer for the wonderful lunch in the Strip in Pittsburgh in 1997 that began our discussion of mechanisms, which soon came to include Carl Craver's useful insights. It has been a delight to participate in this collaboration and to see how it has given rise to such fruitful work by others and us over the last fifteen years. She is grateful to Mark Rollins and others at Washington University in St. Louis for their help during her stay as the Clark-Way-Harrison Visiting Professor of Philosophy, May–June 2008, when the conversations with Carl turned into the beginning chapters of this book. She has had countless fun and enlightening conversations about mechanisms with many colleagues, including Adele Abrahamsen, Garland Allen, Bill Bechtel, Jim Bogen, Pierre-Alain Braillard, Jason Byron, Justin Garson, Stuart Glennan, Christophe Malaterre, Michel Morange, Thomas Pradeu, Bob Richardson, Jim Tabery, Paul Thagard; and with other colleagues about errors and anomalies: Douglas Allchin, Bruce Buchanan, Will Bridewell, and Kevin Elliott,. For many helpful comments on earlier drafts of chapters, she thanks members of the DC-Maryland History and Philosophy of Biology discussion group, including Tudor Baetu, Erin Eaker, Pamela Henson, Sandra Herbert, Erika Milam, Jessica Pfeiffer, Eric Saidel, Lizzie Schechter, and Joan Straumanis. She has enjoyed lively discussions about the mechanism of natural selection with Roberta Millstein, Rob Skipper, Matt Barker, Ben Barros, and Lane DesAutels. Steve Mount of the Center for Bioinformatics and Computational Biology at the University of Maryland helped with biological examples. Lindley owes a special debt to her research associate Nancy Hall, her friend Ben Cranston, and the many graduate and undergraduate students at the University of Maryland who have read and reread various versions of the manuscript and provided helpful comments, including Lucas Dunlap, Mark Engelbert, and Blaine Ford. Lindley also is grateful for the substantial support provided by the University of Maryland during the years of work on this book: a sabbatical leave; the General Research Board of the Graduate School; Jeff Horty and the staff of the Philosophy Department for their support for her and for hosting Carl Craver's visits.

Carl Craver would like to thank Washington University for a sabbatical leave during the 2010–2011 academic year, during which this book came to fruition.

For intellectual contact, Carl is especially grateful to the community with whom he corresponded or worked during this book project: Adele Abrahamsen (UCSD), Garland Allen (Washington University in St. Louis) Colin Allen (University of Indiana), William Bechtel (UCSD), James Bogen (University of Pittsburgh), Dennis Des Chene (Washington University in St. Louis), Frederick Eberhardt (Washington University), Marie Kaiser (Universität zu Köln), David Kaplan (Washington University in St. Louis), Max Kistler (IHPST), James Lennox (University of Pittsburgh), Mariska Leunissen (University of North Carolina), Gualtiero Piccinini (UMSL), Alex Reutlinger (Universität zu Köln), Jim Tabery (University of Utah). And he thanks Pamela Speh for her work on the design and production of the figures and tables, and for the past twenty years of love and support.

He would also like to thank a number of students for assistance at different life stages of this manuscript, especially Faith Steffan, Joshua Borgerding, and Mark Povich. Carl is also deeply grateful to Tamara Casanova, Kimberly Mount, and Mindy Danner for administrative and many other deeply valuable forms of support, and to Louise Gilman and Kaarin Spier for their administrative support during his many stays at the University of Maryland.

We thank Karen Darling, our editor at the University of Chicago Press, for her encouragement and patience over the years that this book came to fruition.

This research has been made possible in part by a collaborative research grant from the US National Endowment for the Humanities: "Because democracy demands wisdom." Any views, findings, conclusions, or recommendations expressed in this work do not necessarily represent those of the National Endowment for the Humanities.

LEARNING HOW CURARE KILLS

In 1804, many years before Darwin boarded the *Beagle*, the explorer Charles Waterton (1782–1865) traveled to British Guiana to take over an estate left to him by his uncle. From there he launched expeditions into the rest of Guiana and into Brazil. He reported his findings in a book, *Wanderings in South America*, and introduced many new types of animals and plants to Great Britain. In one story, Waterton describes joining a bow-hunting expedition for monkeys with some native people.

Having spotted a monkey in a nearby tree, one of the natives draws his bow and fires. The tiny arrow goes wildly off course and lodges in the arm of another member of the hunt. Recognizing that he's been hit, he announces, "Never will I bend this bow again," lays on the ground, and, without further ceremony, dies. The hunter knew, as every member of the party knew, that the tip of the arrow had been dipped in *wouralia*, the local name for a residue made from local plants. And the hunter knew, as every member of the party knew, that once the poison entered his body, nothing could be done to prevent his demise.

Although everyone was quite familiar with the fact that *wouralia* means certain death, no member of the party could say precisely how the plant residue kills its victims. The answer would not begin to appear until nearly forty years later in the 1850s, when Claude Bernard (1813–1878) published his reports describing experiments to discover the mechanism by which that poison, more commonly known as curare, kills its victims. Bernard and his students discovered, for example, that curare does not kill the victim unless it enters the bloodstream and that it does not efficiently do so through the stomach. He determined that curare is a crystalloid and that it can pass through a semi-permeable membrane by dialysis, showing that the poison might be absorbed from the prick of a pin or an arrow. After placing a small piece of curare under a frog's skin, Bernard found that the frog's heart continued to beat long after the frog stopped breathing. He concluded from this that the poison kills the animal by interfering with its ability to breathe. He further showed that if one maintains the frog with artificial respiration, it will eventually live to hop another day.

Bernard began to think that perhaps the poison interferes with the nerves or the muscles required for breathing. He found that curare clearly blocks the ability of motor neurons to cause muscular contractions, but that it leaves the sensory nerves relatively unaffected. After showing that the muscle continues to contract in response to electrical stimulation (that is, that the muscle itself still works), Bernard initially concluded that curare deadened the motor neuron itself. In later publications however Bernard reported that both the muscle and the motor neuron continue to function after the application of curare. The natural conclusion to draw is that curare somehow blocks communication between the motor neuron and the muscle.

Precisely how curare blocks synaptic communication at the neuromuscular junction was not determined until nearly a hundred years after Bernard's pioneering studies in the frog. Neuroscientists have subsequently learned that the neuromuscular junction works through chemical transmission: the electrical signal in the motor nerve causes the nerve to release a chemical neurotransmitter, acetylcholine (ACh). This transmitter diffuses across the gap to the muscle, where it binds to specific receptors for ACh. In the bound state, ACh receptors open to expose a pore through the muscular membrane, allowing passage to charged ions that constitute an electrical signal. Curare acts by interfering with this mechanism. It mimics the shape of ACh, and so binds to the places on the receptor that ACh would occupy, but it does so without opening the channels.

How does curare kill its victims? It enters the bloodstream and makes its way to the neuromuscular junction. There, it blocks chemical transmission from motor neurons, effectively paralyzing the victim. When the diaphragm is paralyzed, the victim cannot breathe, and animals that cannot breathe do not get the oxygen required to maintain basic biological functions. Thus is hunter's lore about the irrevocable effects of curare transformed into scientific knowledge of mechanisms: knowledge of the entities and activities organized together such that they produce, underlie, or maintain the phenomenon in question.

LEARNING TO LOOK FOR MECHANISMS

The search for mechanisms is one of the grand achievements in the history of science. The achievement is first and foremost conceptual: it is the very idea that scientific activity should be organized to advance the discovery of mechanisms that produce, underlie, or maintain the diverse manifest phenomena of our world. The achievement is, second, methodological: it involves the increasing acceptance and refinement of a set of tools for constructing, evaluating, and revising descriptions of mechanisms. No one person created this mechanistic view or the experimental approach characteristic of its champions. In *The New Atlantis*, a work that inspired the social organization of science in the seventeenth century and beyond, Francis Bacon (1561–1626) described a utopian society sustained through the efforts of specialized scientists organized to discover and control nature's hidden causes. During that period, commonly referred to as the Scientific Revolution, thinkers came to see the natural world as a world of mechanisms, just as they came to see science as fundamentally organized around the search for mechanisms. As a consequence, the methods of science came increasingly to be evaluated in terms of their efficiency and reliability as tools in the search for mechanisms. The scientific project, in turn, was justified in many domains by the fact that knowledge of the hidden mechanisms of the natural world offers humans power over the forces of nature that dominate their lives.

Just when and how this mechanistic view of science entered the different fields of biology, specifically, and precisely how the idea of mechanism came to so thoroughly triumph as a way of thinking about explanation in biology, we shall not venture opinions. That it has so triumphed is indisputable. Neuroscientists study the mechanisms of spatial memory, the propagation of action potentials, and the opening and closing of ion channels in the neuronal membrane. Molecular biologists discovered the basic mechanisms of DNA replication and protein synthesis, and they continue to elucidate the myriad mechanisms of gene regulation. Medical researchers probe the genetic basis of cystic fibrosis and how nutrient deficiencies give rise to somatic symptoms. Evolutionary biologists study the mechanism of natural selection and the isolating mechanisms leading to speciation. Ecologists study nutrient cycling mechanisms and the way imbalances in nutrient cycling produce dead zones in places such as the Chesapeake Bay. Across the life sciences the goal is to open black boxes and to learn through experiment and observation which entities and activities are components in a mechanism and how those components are organized together to do something that none of them does in isolation.

Yet there is no tidy story to tell about how this idea took hold in biology. Some of the features of mechanistic biology are discussed in Aristotle's *Parts of Animals*, although we hesitate to call Aristotle (384–322 BC) a mechanist. Certainly, the break from Galen's theories of anatomy during the Renaissance, such as Vesalius's corrections to Galen's human anatomical diagrams and Harvey's demonstration that the blood circulates, share many of the marks of a commitment to the search for mechanisms and of the effort to codify experimental and observational methods for discovering mechanisms. In other respects, however, such theorists were decidedly nonmechanistic, making frequent appeals to Aristotelian notions that would come to be seen as the very antithesis of mechanism in the sixteenth century. René Descartes (1596–1650) imagined a world of small corpuscles colliding with one another and, in *Le Monde*, fashioned innumerable models of mechanisms to explain diverse features of the biological and non-biological world in terms of this basic activity. Yet Descartes famously left room in his world for nonmechanical causal interactions to explain the relationships between minds and brains. Perhaps one could point to the nineteenth century, to Claude Bernard or perhaps to Emil du Bois-Reymond (1818–1896), as powerful figures in part responsible for the stunning extent to which biologists understand what they are doing in terms of the search for mechanisms. Certainly their insistence upon the search for mechanisms fit nicely within the nascent worldview of Charles Darwin (1809–1882), according to which the exquisite adaptedness and diversity of living systems are in fact produced by nothing more than the purposeless mechanism of natural selection.

The mechanical philosophy has been expressed in many ways by many different authors, but one fundamental metaphysical theme is that all phenomena in nature are ultimately explained in terms of a very restricted set of basic, non-occult, non-vital, non-emergent activities. Descartes envisioned the mechanical universe as a billiard-ball universe, made only of things that take up space moving about and clacking into one another. He advanced the bold thesis that everything (except human minds and God) runs by one fundamental activity: movement conserved through collision. It is as if God arranged everything in the world just so and then kicked it. Collision upon collision propagated motion through time, making rivers flow, moving planets about the sun, and sending blood in a circuit about the body. Other mechanical philosophers chose different fundamental, non-occult activities to bottom-out their mechanisms: for some, attraction and repulsion are fundamental; for others, conservation of energy and matter are on the bottom rung. Sometimes the name "the mechanical philosophy" is intended to pick out precisely this kind of austere and materialistic metaphysical commitment.

Although very few contemporary biologists admit to the existence of occult or vital forces in the biological world, it would be false to assume that they embrace anything so austere as these pristine worldviews. Such austere and materialistic forms of the mechanical philosophy are now largely historical curiosities. Contemporary science, let alone biology, is not so restricted in the number and kinds of activities that might appear in its descriptions of mechanisms. The number of acceptable activities, while not unrestricted, is too large to list or count. Some argue that Descartes' austere mechanical philosophy ended once and for all with Newton's introduction of forces and his (reluctant) embrace of action at a distance. Thermodynamic mechanisms were added to the list of activities in the nineteenth century to explain transfer of energy and tendencies to equilibrium. Electromagnetic activities were put on a solid foundation around the same time. Diverse types of chemical bonding discovered in the nineteenth and early twentieth centuries compose biochemical, molecular biological, and metabolic mechanisms. Physiological systems often contain proprietary activities: neurons generate action potentials, hearts and muscles contract, circadian clocks entrain, and the primary somatosensory cortex forms a topographical map of the body. Organisms as a whole engage in diverse kinds of behaviors (such as mating and drinking), and populations of organisms grow, migrate, and divide. The list of non-occult activities acceptable for inclusion in descriptions of biological mechanisms has thus expanded considerably since the austere mechanism of Descartes. Many kinds of activities have been discovered, characterized, and placed on a solid epistemic foundation (using methods that we discuss in later chapters). All these activities are potentially available for inclusion in one's understanding of how a mechanism works.

Whatever the origins of this mechanistic perspective, and however it is related to the austere forms of mechanism that developed in the sixteenth and seventeenth centuries, it is now so thoroughly woven into the fabric of contemporary biology one might easily forget that biology could have taken a different form. Instead of mechanism, one might find a biology of more thoroughly Aristotelian orientation. Such a biology would place particular emphasis on explanations in terms of the goals or functions of organisms, the "that for the sake of which," final, or teleological explanations in Aristotle's four causes. Perhaps it would also emphasize Aristotelian formal causes: this cat has four legs because it has the form of a cat. While mechanists of the sixteenth and seventeenth centuries would allow that considerations of formal and final causes might play a heuristic role in discovering biological mechanisms, they generally insisted that such causes were not to be included among the acceptable explanations for biological phenomena. For the mechanist, the ability to perform a function is a phenomenon that requires a mechanistic explanation; it is not an explanation in its own right. For the mechanist, the form of a cat is either a filler term for mechanisms-we-know-not-what (i.e., the mechanisms producing the various traits and behaviors characteristic of cats) or it is an empty nothing that can do no explanatory work.

Instead of mechanism, biology might have followed the form of the bestiaries of the late Middle Ages, finding in the menagerie of life-forms on earth certain allegorical lessons from a god about how to live our lives. The beaver (*castor* in Latin) instructs us through its namesake behavior to cast off the temptation to sin, lest we suffer. Herbalists might still be studying plants by using their color, for example, to indicate a god's plan for their medicinal effects on humans. For the mechanists, the biological systems of our world are not messages from a creator but fascinating collections of mechanisms that perform the most complicated tasks, as it were, automatically. These systems behave as they do because they are made of components of particular sorts that engage in specific sorts of activities and are organized together such that the organism exhibits its behavior. The mechanistic biologist is not a decoder of theological texts but a kind of curious mechanic, an engineer building a blueprint for the mechanism working in a target system. The goal is not salvation but knowledge of mechanisms and the power that such knowledge promises.

The fact that biology has become a search for mechanisms is not merely a matter of fashion. Biologists look for mechanisms because they serve the three central aims of science: prediction, explanation, and control. First, knowing the mechanism usually allows one to predict how the phenomenon will behave. If one knows how a mechanism works, one can say how it would work if it were placed in different conditions or given different inputs. Second, and related, describing the mechanism for a phenomenon serves to explain the phenomenon. In some cases one can literally *see* how the mechanism works from beginning to end. Finally, knowing the mechanism potentially allows one to intervene into the mechanism in order to produce, eliminate, or change the phenomenon of interest. Biological mechanisms, in other words, are of interest because we want to bring them under our control: for production (as in agriculture and farming), for healing (for the purposes of medicine and pharmacology), and for environmental management and protection (in ecology). Examples abound: One's understanding of how a normal mechanism fails in disease can guide one in the search for cures and preventions. One's understanding of how a natural ecosystem works might suggest interventions to control or ameliorate the effects of an invasive species.

In saying that the search for mechanisms has come to dominate biological research, we do not intend to suggest that biology is *only* concerned with the discovery of mechanisms. Some biological research programs are not particularly driven by the search for mechanisms. A taxonomist might be interested in cataloguing the diversity of life in an Amazonian rain forest with no interest in mechanisms whatsoever. An epidemiologist might be interested in modeling or predicting how a disease will spread in a population without knowing the precise mechanisms by which the disease spreads. A developmental psychiatrist might be interested in the frequency with which individuals with autism fail to understand what (or that) other people are thinking without a concern for why individuals with autism have these characteristics. A molecular neurobiologist might be interested in the structure of an ion channel with no regard to the mechanisms by which the channel opens and closes, just as an anatomist can (in principal at least) study how parts are arranged spatially within the body independently of any consideration about what those parts do and how they are organized together. Categorization, generalization, modeling, observation, and prediction are, in fact, often useful in the search for mechanisms, but the value of these scientific practices is not exhausted by their contribution to the search for mechanisms. Science, in short, is not defined as the search for mechanisms; still, much of biology is in fact driven by the search for mechanisms. This is a detail worthy of our sustained, isolated attention. For this reason we focus here on the search for mechanisms, and we discuss these other practices only in so far as they contribute to that search.

In Chapter 2 we say what mechanisms are, offering an abstract characterization of the main features of any given mechanism. The concern here is with describing the causal structures, the mechanisms, that are the focus of biological explanation. In Chapter 3 we focus instead on descriptions of mechanisms, the drawings, diagrams, equations, and models that scientists use to represent how a mechanism works. We distinguish mechanism schemas, which are abstract descriptions of a mechanism with placeholders that can be filled in with known entities and activities, from mechanism sketches, which are incomplete schemas. Mechanism schemas and mechanism sketches are both hypotheses about how a given mechanism works: they are hypothesized mechanisms for the phenomenon.

The primary point of these preliminary chapters is to frame the target of the search for mechanisms: to build a schema that adequately captures the relevant features of a mechanism. That scientists are searching for *mechanisms* rather than mere correlations, final causes, formal causes, or divine purposes, places constraints on what will properly terminate that search. This introductory discussion about what mechanisms are and how they are described anchors the view of discovery to be developed in subsequent chapters. The view of discovery includes both a descriptive model of the mechanism discovery process and a set of questions, constraints, and strategies that might be used by scientists at different stages of that discovery process. The product of the search for mechanisms shapes the process by which it is discovered.

DISCOVERY: THE PRODUCT SHAPES THE PROCESS OF DISCOVERY

We distinguish four different stages of the mechanism discovery process: characterizing the phenomenon (Chapter 4), constructing a schema (Chapter 5), evaluating the schema (Chapters 6, 7, and 8), and revising the schema (Chapter 9). Although we will often describe these components as stages in mechanism discovery, we emphasize that they are often pursued in parallel and in interaction with one another. A central feature of our view is that mechanism discovery is frequently a piecemeal and protracted affair. It is piecemeal in the sense that one might work on a part of a mechanism, or an aspect of its function, while leaving much else about the mechanism inside a black box. It is also piecemeal in the sense that the different stages of discovery frequently interact with one another: one is forced to recharacterize the phenomenon in the face of learning about the mechanism, or one is forced to reevaluate experimental findings because one recognizes a previously unrecognized region of the space of possible mechanisms. We describe these components of discovery as stages only for terminological convenience.

The first stage, characterizing the phenomenon, frames the problem to be solved in the search for mechanisms. This stage involves developing a more or less precise description of the behavior or product of the mechanism as a whole. In Chapter 4 we discuss the diverse kinds of conditions included in the characterization of a complex phenomenon. The nature of the phenomenon often provides clues about the kind of underlying mechanism that might possibly be responsible for it. In protracted discovery episodes, scientists are often forced to reconceptualize the phenomenon and start the search for mechanisms anew. We discuss some signs that the time has come to recharacterize the phenomenon and some possible ways of revising that characterization when such revision becomes necessary.

The second stage of mechanism discovery, constructing a schema, involves generating a space of possible mechanisms for a given phenomenon. All of the strategies we discuss in this chapter rely on using the conceptual resources and background knowledge potentially available to the scientist facing the discovery problem. Sometimes, for example, scientists reason by analogy in generating possible mechanism schemas. Other times they borrow a schema from a neighboring field. Sometimes they assemble a novel kind of schema by cobbling together modules known to be repeated in the biological world. Such strategies rely crucially upon the scientist's being equipped with an appropriate store of analogues, schemas, and modules. Basic science training is designed to equip students with this store of biological components and mechanism-types; the broader one's training, the more conceptual resources one can bring to bear upon a particular discovery episode. In Chapter 5 we discuss resources that one might consult for constructing new schemas.

The third stage of discovery, evaluation, involves sorting good from bad mechanism schemas. We describe this process metaphorically as a matter of revealing the empirical and conceptual constraints on the space of possible mechanisms for a given phenomenon. At the most abstract level mechanism schemas are evaluated in terms of their depth, their completeness, and their correctness. We discuss these dimensions of evaluation and some tests for evaluating a schema along these dimensions in Chapter 6. Our orientation toward these questions might be described as a garden-variety realism moderated by a sensible pragmatism. The very idea that mechanism schemas are evaluated in terms of their *correctness* presumes that there is a fully instantiated target mechanism in the world and that one can assess the degree of fit or mapping between items in the schema and items in the target mechanism. The very idea that mechanism schemas are evaluated in terms of their *completeness* presumes that there is a complete target mechanism in the world and that one can assess the extent to which the schema includes all and only the entities, activities, and organizational features in the target. In speaking this way, we also intend to allow that a schema can be complete and correct *enough* for the purposes at hand without being fully complete or correct. One can acknowledge the ideals of completeness and correctness for descriptions of mechanisms while, at the same time, recognizing that science often traffics in idealized and incomplete schemas.

Empirical constraints guide the search through the space of possible mechanisms. In Chapters 7 and 8 we discuss different kinds of constraints on mechanism schemas, and we show how they can be used to structure the space of possible mechanisms, occluding some portions of that space and revealing others for consideration. The material structure of a mechanism—that it is composed of parts with sizes, shapes, and structures, that it has a particular temporal sequence, that it is composed of one kind of entity and not another or makes use of one kind of activity and not another—directly contributes to the kinds of evidence by which scientists evaluate mechanism schemas and to the cognitive strategies by which scientists reason about how a mechanism works.

The contents of Chapters 7 and 8 map, somewhat imperfectly, onto an imperfect distinction between constraint-based reasoning about mechanisms and experimental reasoning about mechanisms. An experiment, in the narrow sense intended here, involves intervening to change some part of a mechanism or some background condition in order to learn something about how the mechanism works. Constraint-based reasoning is grounded in evidence that does not involve such intervention. Findings concerning the spatial and the temporal organization of a mechanism are a good example of constraint-based evidence that shapes the space of possibilities. Sizes, shapes, locations, positions, orientations, conformations, and connections are some of the spatial constraints. Orders, rates, and durations are some of the temporal constraints. In Chapter 7 we use William Harvey's extended mechanistic argument for the circulation of the blood as an example of such constraint-based reasoning. In Chapter 8 we analyze the structure of some classic experiments in biology, with particular attention to experiments directly used to address special kinds of questions that commonly arise as one tries to learn how a mechanism works.

The final stage of mechanism discovery is revision. Revision is required when an empirical anomaly raises an apparent challenge to a favored mechanism schema. In Chapter 9 we survey a range of choices that the scientist might consider in response to an anomaly for this mechanism schema. Some of the choices concern whether the apparent challenge to the schema can be cordoned off as, for example, a failed experiment or as beyond the scope of the intended schema. When such cordoning off fails, and the anomaly is determined to directly challenge a schema, there are a number of search strategies that scientists may use to localize the site of fault within their schemas and to correct them in response.

Many people are simply unaware of or are ill equipped to appreciate how the massive body of biological knowledge concerning the mechanisms of the living world is grounded in empirical facts and cobbled together over time. And if one cannot appreciate this protracted and continuing discovery project, this project of discovering mechanisms, then one cannot fully appreciate why the working hypotheses of contemporary biology are worthy of our respect. One cannot appreciate why the contemporary version of Darwin's theory of evolution by natural selection is an especially good and well-certified theory. One cannot appreciate the Mendelian geneticist's understanding of heredity or the significance of the discovery of DNA for understanding how life works. One cannot understand biology or its history without understanding these protracted discovery episodes. And as most of these episodes are driven by the goal of discovering mechanisms, one cannot understand biology without understanding how phenomena are characterized and how mechanism schemas are constructed, evaluated, and revised. One can read this book at one level as an effort to characterize this process.

STRATEGIES FROM HISTORY

This book can also be read as a piece of compiled hindsight about diverse questions, constraints, and strategies for discovering mechanisms. As such, it is something of a guidebook to fruitful reasoning strategies that scientists can use in the search for mechanisms. We have tried to codify and distill the common wisdom that scientists exhibit when they discover mechanisms. Such wisdom is the hard-fought product of the scientists' efforts to learn how best to learn about the hidden structure of the world. By watching this discovery process in action, we gain some insight into the fruitful questions to ask and to the available moves in the course of such a protracted discovery episode. One of our goals is to abstract from the details of specific cases to formulate general strategies that together form a tool kit for discovering a mechanism, that is, for formulating and addressing central questions in the discovery of mechanisms.

We are aware of areas in the philosophy of science that emphasize formal discovery methods by which, for example, causal models can be inferred on the basis of correlational data or from data about the effects of interventions on observed statistical dependency relations. In a few places, especially in our discussion of experiments for testing causal relevance, we draw upon some of the insights of this formal literature. Yet we are after something different and, in a way, more basic to our reasoning about mechanisms, something that goes beyond mere patterns of statistical dependency among variables. Such reasoning, we emphasize repeatedly, turns on the embodied nature of mechanisms: that they are not mere correlations among variables but entities and activities with spatial and temporal properties organized to produce, underlie, or maintain a phenomenon. The entities have sizes, shapes, positions, and orientations. The activities have orders, rates, and durations. The items that appear in mechanism schemas are not merely defined variables but variables that stand for entities and activities about which we have a great deal of background knowledge. Scientists draw on that background knowledge as they construct, evaluate, and revise mechanism schemas. They reason about function on the basis of structure; they reason to likely causes from facts about effects; they restrict the space of possible mechanisms to just those with components known or likely to be found in the system in question.

As we discuss the four aspects or stages of discovery (characterization of the phenomenon, construction, evaluation, and revision), we compile crucial questions, constraints, and strategies to aid in solving diverse discovery problems. For example, in Chapter 4 we discuss how to characterize a phenomenon so as to maximally constrain the space of possible mechanisms. We discuss several ways that one might mischaracterize the phenomenon as well as some strategies for re-characterizing the nature of the phenomenon in response to those failures. In Chapter 6 we discuss some conceptual strategies for testing the completeness and correctness of one's mechanism schema. In Chapter 7, we discuss the different kinds of empirical constraints that have proved especially useful in the discovery of mechanisms, and we show how different kinds of constraints shape the space of possible mechanisms. In Chapter 8 we discuss experimental strategies for testing mechanisms, including experiments for testing causal relevance, experiments for finding components, and experiments for filling black boxes in mechanisms. In Chapter 9 we discuss strategies for dealing with empirical anomalies for a mechanism schema, and, ultimately, for revising the schema in light of those anomalies.

Perhaps this compiled hindsight from the history of biology will help a biology student to see how a particular piece of research fits into a wider theoretical context. Perhaps this set of strategies will help a puzzled scientist to think about how to construct, evaluate, or revise a mechanism schema. Perhaps this historical study will help the curious to refine their skills at learning for themselves how things work. And perhaps, short of that, our discussion will give one a new appreciation for how our understanding of biological mechanisms is placed on a firm evidential foundation.

THE MECHANISTIC INTEGRATION OF BIOLOGY

Biology is a collection of fields and subfields. A field is unified by a common problem or set of questions, a shared vocabulary for describing the items in that domain, a set of experimental and observational tools to use for answering those questions, a set of accepted protocols for using those tools, and, finally, a set of concepts and theories in terms of which the problem might be solved. A small sample of fields from biology would include, for example: anatomy, physiology, cytology, ecology, embryology, epidemiology, genetics, neuroscience, evolutionary biology, botany, and zoology. One might be forgiven for thinking of these fields as patchwork windows on the biological realm that overlap in various ways but also sometimes operate in considerable isolation from one another. One striking subtext of the numerous examples of mechanism discovery we consider is the frequency with which exciting discoveries about mechanisms requires collaboration among researchers in different fields of biology. In each case, the search for mechanisms provides a scaffold around which the findings of different scientific fields are integrated in a common explanatory objective. In Chapter 10 we discuss the way the search for mechanisms provides a framework for integrating the contributions of diverse fields of biology, across multiple levels, and through temporal series.

THE PRAGMATIC POWER OF MECHANISM

Of the many forms biology might have taken, biology has in fact taken a mechanistic turn. This turn involves a commitment to the discovery of mechanisms as a fundamental goal. This turn also involves the acceptance of a set of constraints that must be satisfied by any acceptable description of a mechanism and a set of investigative techniques designed to bring those constraints into view. Our goal is to make the central aspects of this continuing mechanistic perspective explicit with the aims, first, of clearly expressing something central to the structure of contemporary biology, and second, of offering a set of questions, constraints, and strategies that might help people to undertake the search for mechanisms.

Biology has taken a mechanistic turn in large part because of the instrumental power of mechanistic knowledge. Setting aside for a moment the measure of understanding supplied when one knows how something works, mechanistic knowledge is also of tremendous use for prediction and control of phenomena in the domain of biology. When we know how something works, we are often in a position to predict how it will likely behave in the future, perhaps even under novel circumstances. When we know how something works, we know various ways to intervene and make the mechanism work for us. We are better able to diagnose its failures, and we generate ideas about how to fix it. We learn design principles that might be used to build artifacts or prostheses. We apply our mechanistic knowledge to design new experimental techniques that allow us to address more and more precise questions in our experiments. We sometimes learn to fine-tune a mechanism and make it work the way we want it to work. The mechanistic knowledge supplied by contemporary biological research thus continues to promise (and in many stunning cases, deliver) control over nature, for good and for ill. In its optimism about the fruits of mechanistic knowledge, contemporary biological science carries the torch for a core component of Bacon's image of science. We discuss the connection between mechanisms and control in Chapter 11.

BIBLIOGRAPHIC DISCUSSION

Bernard began his classic work on curare early in his career, with the first major publications on this topic appearing in the early 1850s (1850a; 1850b; 1851a; 1851b). He is best known for his review of several such poisons in 1864. Our historical discussion makes use of Lee's (2005) helpful review of the history of curare, as well as Birmingham's (1999) and Black's (1999) discussions of Waterton. The puzzle of curare's mechanism of action was resolved finally (or so we assume) by Del Castillo and Katz (1957). For more on the history of chemical neurotransmission, see Elliot Valenstein's *The War of the Soups and the Sparks* (2005).

Westfall (1971) provides a very readable history of the origins of the mechanistic worldview. For more on Descartes' and late Cartesian versions of the mechanical philosophy and its context within the new science, see Boas (1952), Des Chene (2001), Dijksterhuis (1961), Gabbey (1985; 1990), Garber (1992), Shapin (1996) and Wilson (1999). For more on du Bois-Reymond, see his 1848 work on animal electricity, as well as the discussion in Cranefield (1957). On the varieties of the mechanistic perspective and the opposition between mechanism and vitalism in nineteenth- and twentieth-century biology, see Allen (2005).

We discuss how contemporary views of mechanism differ from those of the seventeenth century in Craver and Darden (2005). For more on the historical discovery of diverse kinds of activities, see the brief discussion in Section 1.5.2 in Machamer, Darden, Craver (hereafter MDC 2000). Darden (1987) proposes using historical cases to produce compiled hindsight. Technical books and articles on our views of discovery include Darden (1991; 2006), Craver and Darden (2001), Darden and Craver (2002); and Craver (2007) on mechanistic explanation.

Our view about the mechanistic structure of biological sciences is compatible with and complementary to more formal approaches to thinking about causal inference (see Pearl 2000; Spirtes et al. 2001). Our focus in this book is on qualitative reasoning strategies in the search for mechanisms. It is in our view an interesting and open research question whether and how such qualitative approaches might be compared and combined with formal systems for describing and regimenting causal inference. For a preliminary investigation, see Newsome (2003).

We are grateful to Robert Olby for sharing his unpublished work on bestiaries, to Peter Machamer for insights into the role of Bacon's vision in the development of the mechanical philosophy, and to Ted McGuire for discussions of the epistemological and metaphysical features of the scientific worldview that ascended in the seventeenth century.

2 BIOLOGICAL MECHANISMS

WHAT IS A MECHANISM?

Mechanisms are how things work, and in learning how things work we learn ways to do work with them. Biologists try to discover mechanisms because mechanisms are important for prediction, explanation, and control.

Biologists seek mechanisms that produce, underlie, or maintain a phenomenon. Stated most generally:

> *Mechanisms are entities and activities organized such that they are productive of regular changes from start or set-up to finish or termination conditions.*

Below we elaborate upon the different parts of this characterization, one at a time, to bring the target of the search for biological mechanisms more clearly into view. As we elaborate throughout the book, the intended target of the search—mechanisms—shapes the process by which biologists try to find them.

MACHINES VERSUS MECHANISMS

It is often easy to understand how human-made machines work. One gear turns, and its teeth mesh with another gear, which turns too. This was an appeal of the seventeenth-century clockwork universe: it was understandable, given the analogies to machines of the day. And sometimes it is also easy to understand biological mechanisms; they have parts and are driven by activities that are machinelike. A DNA double helix opens like a zipper and its bases mesh with other bases that have complementary shapes (and charges).

However, biological mechanisms are often quite unlike machines. A machine is a contrivance, with preexisting, organized, and interconnected parts. Paradigm examples are the mechanical clock, the water pump, the internal combustion engine, and the computer. Biological mechanisms have been tinkered together under mutual constraints through evolution by natural selection and through development. Their parts may be synthesized on the fly and rapidly degraded or they may be more stable. But not all the parts are necessarily in place prior to the operation of the mechanism. The blueprint of the typical biological mechanism is decidedly messier than the blueprint for even complicated machines.

Yet the difference between mechanisms and machines is not merely a matter of their complexity and tidiness. Mechanisms are characteristically active; they are how things work, when they work. Machines exist in active and inactive states. A stopped clock is a machine but not a mechanism. It is an organized assemblage of parts without any activities. Biological mechanisms do things. They move things. They change things. They synthesize things. They transmit things. They may even hold things steady. Biologists investigate mechanisms because they hope to understand how things work and, as a consequence, to understand how to control the activities of biological mechanisms. This cannot be done without understanding the distinctive activities found in biological mechanisms.

Finally, and perhaps most to the point, one and the same machine might be composed of a number of wholly distinct mechanisms. The car has a windshield wiper mechanism and an engine. It also has a radio. The windshield wiper and the engine are not both part of a single mechanism. Each is part of a different mechanism, for a different phenomenon, in one and the same machine. Likewise, an organism is not a mechanism but a collection of mechanisms colocalized in a body. Although the idea of a biological mechanism no doubt has some historical connection with the idea of a mechanical contrivance, the idea of a mechanism flourishing in contemporary biology is at most analogous to the idea of a machine. The terms mechanism and machine are not synonymous.

COMPONENTS AND FEATURES OF MECHANISMS

Let us return now to the main working parts of our characterization of biological mechanisms. These are summarized in Table 2.1.

ENTITIES AND ACTIVITIES

Mechanisms are composed of both *entities* and *activities*. The entities are the parts in the mechanism with their various properties. Activities are the things that the entities do; they are the producers of change. An enzyme (entity) phosphorylates (activity) a protein (entity). A neuron (entity) releases (activity) a neurotransmitter (entity). Organisms, cells, and macromolecules are kinds of entities. Pushing, pulling, bonding, diffusing, opening, blocking are kinds of activities. Entities have properties, such as structure and orientation, that enable them to engage in specific activities. Activities, correlatively, require the existence of certain entities with specific properties. Diffusion across a membrane, for example, requires entities of appropriate sizes, concentration differences on each side of the membrane, and a membrane permeable to the relevant ions. Mechanisms are collections of entities and activities organized in series, joins, or circles so that they do something that the components cannot do on their own.

Components and Features of Mechanisms
Entities and Activities
Setup, Start, and Finish Conditions
Productive Continuity
Regularity
Organization
Spatial
Temporal
Active
Levels of Mechanisms
Topping-Off
Bottoming-Out
Mechanistic Context

Table 2.1

In descriptions of mechanisms, nouns (channel, terminal, enzyme) usually refer to entities. Active verbs (bond, release, phosphorylate) usually refer to activities. Entities are identified by their properties, by spatiotemporal locations, by boundaries (e.g., a surrounding membrane), by subparts (for example, parts that are chemically bonded to each other and not bonded to parts outside the entity, such as macromolecules), by parts that have their own integrity (for example, organisms as parts of populations), and by their durations (e.g., stable over generations, hours, minutes, or milliseconds.). Similarly, activities are identified by their spatiotemporal location, rate, and duration, by the types of entities that can engage in them, by the types of properties that make them possible, and by the startup conditions that enable them. More specifically, activities are distinguished from one another by their mode of operation (e.g., contact action versus attraction at a distance), polarity (e.g., attraction, repulsion, or both attraction and repulsion), energy requirements (e.g., how much energy is required to form or break a chemical bond), and range (e.g., electromagnetic forces have a wider influence than do the forces in the nuclei of atoms). The goal in describing a mechanism is, in part, the goal of finding the component entities and activities by which the mechanism works.

Consider for example the mechanism of DNA replication. The double helix of DNA (an entity) unwinds (an activity) and new component parts (entities) bond (an activity) to both parts of the unwound DNA helix. DNA is a nucleic acid composed of several subparts: a sugar-phosphate backbone and nucleic-acid bases. As DNA unwinds, the bases exhibit weak charges, properties that result from slight asymmetries in the molecules. These weak charges allow a given

DNA base and its complement to engage in the activity of forming hydrogen (weak polar) chemical bonds; the specificity of this activity is due to the geometric arrangements of the weak polar charges in the subparts of the base. In short, entities with polar charges enable the activity of hydrogen bonding. After a complementary base and other components align, then the DNA backbone forms via stronger covalent bonding. The mechanism proceeds with unwinding and bonding together (activities) new parts, to produce two helices (newly formed entities) that are, more or less faithful, copies of the parent helix.

Nucleic acids and proteins are among the *working entities* of molecular biological and biochemical mechanisms, such as the mechanisms of DNA replication and protein synthesis. Working entities are entities engaged in the operation of the mechanism. Working entities have the requisite sizes, shapes, structures, charge distributions, or other activity-enabling properties to play the required roles in these molecular mechanisms. Working entities may have localized active sites, such as the portion of the protein that serves as the site for the docking of a substrate to an enzyme. Alternatively, active sites may be distributed throughout the entity, as the slightly charged bases are distributed along the unwound DNA double helix. Of course, the double helix has subparts, such as the protons and neutrons in the atoms that compose the macromolecule; however, protons and neutrons are not working entities in the mechanism of DNA replication.

SET UP, START, AND FINISH CONDITIONS

It is an idealization to describe mechanisms as working from start to finish. Not all mechanisms work like that. Some mechanisms may have a privileged endpoint, such as rest, equilibrium, neutralization of charge, a repressed or activated state, elimination of something, or the production of a product. Such *termination conditions* might be reached because, for example, some reactant has been used up. Talk of just a single input or output is in many cases too impoverished to capture the wide variety of conditions that enable the first stage of the mechanism to proceed or the wide variety of conditions that may be considered its termination state.

The idea that mechanisms work from start to finish applies best to mechanisms that are *linear* from beginning to end, with the first stage giving rise to the second, the second giving rise to the third, and so on. We explain how a protein is produced, for example, starting with DNA and ending with the peptide bonds between ordered amino acids. Not all mechanisms are linear in this sense. Some mechanisms are organized in *cycles* or have modules that work in cycles. The Krebs cycle (one of several steps in the metabolism of sugar) is typically drawn in the form of a circle with its various products exiting the mechanism at key junctures and with its residues available for reuse in the next stage of the cycle. Some mechanisms are more clearly described as *underlying* a phenomenon rather than *producing* it from some earlier state. The mechanism of the action potential (a wavelike electrical potential charge that travels along the membrane of the nerve cell) underlies or implements the phenomenon of the action potential; it does not produce it. This mechanism involves the choreographed opening and closing of sodium- and potassium-channel proteins and the diffusion of ions across the nerve cell's membrane. Other kinds of mechanisms are best described as *maintaining* an equilibrium state, starting from any of a number of possible points of disequilibrium and ending in just one or a few termination states. For example, blood sugar levels are maintained in part by mechanisms that regulate insulin production. Clearly it is a mistake to assume that all mechanisms have a linear organization; in biology, few mechanisms satisfy this assumption.

Some mechanisms do not have any state that could be described as a trigger from the external environment constituting a start stage. Their behavior might better be understood as internally generated, or constitutively produced, without an environmental trigger. Likewise, a mechanism that operates perpetually (until death) doesn't exactly have a termination condition, although it does have an end in producing, underlying, or maintaining something.

PRODUCTIVE CONTINUITY

Mechanisms are productive in that they are how some end state, product, process or change comes about. To show how a mechanism works, one shows how an earlier stage gives rise to the next. Each stage of the mechanism makes a difference to what happens at one or more subsequent stages. This is what it means for one stage to produce another: each stage drives, makes, allows, inhibits, or prevents its successor. An ideal in describing a mechanism is to reveal the productive continuity of the mechanism, eliminating gaps or black boxes by specifying the relevant entities and activities, thereby filling the space between the beginning and the end of the mechanism.

REGULARITY

Most interesting biological mechanisms are *regular* in the sense that they usually work in the same or a similar way under the same or similar conditions. They are assumed in this sense to be deterministic. For example, the mechanism of DNA replication begins with one double helix and ends with two. One double helix unwinds, and each half provides a template along which complementary bases are aligned, yielding two helices at the end. Crucially, however, the two helices may be more or less like the parent one, depending on how reliably the copying mechanism operates and the fidelity of repair mechanisms in checking for and correcting errors. In some species of bacteria, for example, the amount of regularity can be tuned up or down, depending on environmental conditions. In more stressful environments, what is appropriately called the SOS mechanism produces copies of the DNA with more mutations in the daughter helices by utilizing a more error-prone enzyme for copying the DNA. With a bit of luck, some mutants will be able to withstand the challenging conditions.

Although the term mechanism has historical associations with determinism, that association is misleading for thinking about mechanisms in contemporary biology. Many mechanisms are stochastic in the sense that *indistinguishable* conditions, given the state of knowledge and technology, can give rise to very different end states. For example, when the action potential arrives at the axon terminal, one, two, or more quanta of neurotransmitters may be released, or none at all. One can generate frequencies at which each of the outcomes will occur, but one cannot predict with certainty on this basis just which outcome it will be in a particular case. The mechanism of natural selection has a tendency to result in the better-engineered variant organism leaving more offspring in the next generation, but interfering conditions abound. So to say that a mechanism is regular is not to say that it is deterministic. Determinism is only the limit of regularity in most biological mechanisms.

ORGANIZATION

The entities and activities in a mechanism are *organized* spatially, temporally, and actively such that they produce the phenomenon. *Spatial* organization includes such things as the locations, sizes, shapes, and orientations of component entities. *Temporal* organization includes the orders, rates, and durations of the stages. Components with a different spatial and temporal organization often act differently. *Active* organization includes facts about which components make a difference to which others and how such differences are made (that is, by which activities).

Active organization distinguishes mechanisms from mere aggregates (or heaps) of matter, such as piles of sand. The parts act and interact with one another in such a way that the whole is literally not a mere sum of its parts. Mechanisms are in this sense nonaggregative: the parts of the mechanism are organized in ways that go beyond, e.g., the contribution made by the mass of a grain of sand to the mass of the pile. Mechanisms are not mere sums of the properties of their component parts; their components are spatially, temporally, and actively organized.

LEVELS OF MECHANISMS

Biological mechanisms typically span multiple levels. Scientists working at higher levels work on ecosystems, populations, and the behaviors of organisms within their environments. Others study mechanisms within organisms, such as the nervous or circulatory systems, and mechanisms within organs and cells, and, ultimately, mechanisms with smaller entities, such as macromolecules, small molecules, and ions. Different fields of biology are often (to a first approximation) associated with different levels. Molecular biologists study molecules. Cell biologists study cells. Electrophysiologists study channels in cell membranes, cells, and networks of cells. Ethologists study animal behavior. Ecologists study populations of organisms and ecosystems.

This informal and popular notion of levels—*size levels*, we'll call them—in many ways misrepresents the structure of biological science. This perspective emphasizes and reinforces a blinkered and compartmentalized perspective on how biological research is and ought to be organized. It misrepresents biology because many discoveries in biology involve the collaboration across different fields that work at roughly the same size level. A single-unit electrophysiologist might collaborate with a cellular-level anatomist to study a physiological wiring diagram just as a molecular biologist and a biochemist might each contribute distinctive expertise to the understanding of protein synthesis. The idea of size levels emphasizes the differences between fields (that they describe different objects, use different vocabularies, wield different techniques, etc.) over the common, interfield project of discovering a mechanism and learning how it works.

In such an interfield project, researchers often direct their attention to different components in a hierarchy of levels of mechanism. Most biological mechanisms are decomposable into lower-level components and activities, which are themselves decomposable into lower-level components and activities, and so on through the hierarchy.

Levels of mechanisms are thus defined in terms of the relationship between the behavior of a mechanism as a whole and the behavior of a component in that mechanism. Each step in this decomposition hierarchy adds another level in the description of the mechanism. To fully understand a food chain, for example, one has to understand the relationships among many different items in an ecosystem: the behaviors of individual organisms, the physiological function of biological systems within organisms that mediate those behaviors, the operations of organs composing those systems, the activities of cells within them, the bonding and breaking of molecules into which the food is decomposed, and the tireless working of synthesis mechanisms for forming flesh and blood and stems and sap.

Multifield interactions and multilevel mechanisms are signs of healthy research strategies, and the artificial division of biology into size levels of investigation encourages one to think in terms of boundaries between fields rather than the exciting project of building schemas that integrate multiple levels of mechanisms and bring the critical eye of multiple independent fields to bear upon one and the same phenomenon or mechanism. Levels of mechanisms, in contrast, come into view only as part of an integrative project of bridging across different fields of research and aspects of biological phenomena. Interfield interactions of this sort are often crucial in discovery episodes.

Descriptions of multilevel mechanisms must necessarily *top off* in some highest-level phenomenon of interest and *bottom out* in some lowest-level mechanisms beneath which more detail is either superfluous or unavailable. Which level is the most important for understanding a phenomenon, and how many levels a description of a mechanism must have are partly empirical questions and partly practical questions. Where the mechanism tops off depends on one's most fundamental explanatory question. Structural chemists interested in the primary, secondary, and tertiary structures of proteins sometimes have no interest in the higher-level mechanisms to which those proteins contribute. Biologists, in contrast, are interested in particular proteins precisely because they play a crucial role in a higher-level system, such as a cell signaling mechanism or a developmental mechanism of some sort. It is an empirical question whether the mechanism or component in question plays a role in that higher-level mechanism; it is at least partly a practical matter whether that higher-level mechanism is interesting and worthy of investigation.

The bottom-out point for a description of a mechanism might be limited simply by the knowledge of a science at a time. It might be that the mechanisms beneath a certain level of description are unknown. It might be that no techniques are available to detect or manipulate phenomena at that level. However, it might also be that further facts about still lower-level mechanisms are irrelevant to the explanatory project in which one is engaged. If one is interested in why sickle cell anemia persists in populations, one will have to attend to population genetics and to the selection pressures acting on heterozygotes and homozygotes for the relevant allele in geographical regions where malaria is present. If one is interested in designing drugs to treat sickle cell anemia, however, one might be more interested in cellular and molecular aspects of the mechanism. What counts as the bottom out level, in other words, depends on what one is trying to accomplish.

FUNCTIONS

It is often useful to describe the components of multilevel mechanisms in terms of their functions, that is, in terms of their contribution to the behavior of the mechanism as a whole. The circulatory system is a mechanism that can be described at multiple levels. The activities of the circulatory system are explained in part by the heart's pumping of blood, the kidney's filtration of the blood, and the venous valves' regulation of the blood's direction of flow. These component activities can themselves be described in terms of their underlying mechanisms: the heart's pumping can be explained by the contractions of the heart muscles, the behavior of individual cells, their ion channels, and their membranes. Similarly, the kidneys' filtration can be explained by the organized activities of its component glomeruli, pores, tubules, and ionic gradients. The explanation for each could continue further still into the molecular mechanisms that allow these parts to work.

A given part of a multilevel mechanism can be described at its next higher level, its own level, or at a lower level, which we'll designate as the +1 (or +n), 0, or −1 (or −n) mechanism level. To describe an item functionally is to describe it contextually in terms of the contribution that it makes to some higher-level mechanism. The description either explicitly or implicitly describes how a given part is organized together with other components in the mechanism. The amount of context included in such a functional description of an item's higher level can vary considerably depending on one's descriptive purposes. The heart might be said to distribute oxygen and calories to the body, to circulate the blood, to expel the blood, and to contract. (To find the function, one might continue to a +2 or, in general, a +n level. For example, one might discuss the function of the heart in preserving the life of the organism or the function of the heart in vertebrates compared to other groups that lack a specific pumping organ.)

Alternatively, one might describe the heart at its own (0) level as contracting; that is something that it does in isolation, described as if it has no help from other components in the circulatory system. Functional descriptions focus on things that the heart cannot do by itself without being organized together with other entities and/or activities in a higher level. The heart cannot distribute oxygen and calories to the body without oxygen, calories, and a body. Nor can the heart circulate the blood if the veins and arteries are not connected in a circuit.

Typically, a given component will be part of many higher-level mechanisms. The heart makes *glub-blub* noises and generates heat, but these are not relevant to the circulation of the blood, even if they may be relevant to other containing mechanisms in which we might be interested. Heart noises play a functional role in a diagnostic context (a different higher-level mechanism), allowing a physician, for example, to diagnose a mitral valve prolapse. The generation of heat by the movements of the heart plays a limited functional role in the different higher-level mechanism of thermoregulation. Which higher-level mechanism is relevant depends on the context in which the mechanism is being discussed and the chosen phenomenon of interest. Describing an item's role function is a perspectival affair. That is, it depends on the higher-level perspective from which one views the component's functional role.

Some biologists and philosophers use the term function to describe a more restricted class of *proper* functions. The proper function of a part is the role the part played in the evolutionary history of the organisms that have it. One might claim that the proper function of the heart is to pump blood, if it can be shown (plausibly) that the pumping action of the heart contributed to its persistence across generations as a biological trait. The idea of a functional role in a mechanism, however, is not restricted to behaviors of a component that contribute or have contributed (in the recent or distant past) to survival and reproduction. One could, for example, talk of the functional role of a component in a disease mechanism even if the component did not play a historical role in the persistence of the disease. The term role captures this usage.

So understood, mechanistic and functional descriptions are two sides of the same coin. To describe a component functionally is to describe how that component is integrated into a higher-level mechanism. To describe a component in terms of its underlying mechanisms, however, is to show how the isolated behavior of that component (its own o level) is integrated with lower-level mechanisms. Functional and decompositional characterizations at various topping-off and bottoming-out levels play opposing, yet crucial, roles in the integration of multilevel mechanisms.

MECHANISMS IN A WIDER CONTEXT

Mechanisms are located in wider contexts. In order to situate a mechanism in a wider temporal context, one seeks mechanisms that occur before it and provide some of its set up or start conditions. In an evolutionary context, the origin (or origins) of life began a temporal chain of biological mechanisms extending to the present. In an embryological context, many mechanisms began with a sperm and egg uniting to produce a zygote from which the adult organism develops. In physiological contexts, a mechanism may begin with ingestion of a foodstuff, which is decomposed to yield energy (in the form of ATP molecules) and more elemental molecular components (such as amino acids) that can be built up into body parts. In ecological food chains, an energy source (usually the sun, but also the heat from thermal vents in the deep ocean) begins the mechanisms by which primary producers such as plants begin the production of food stuffs consumed by others downstream in the food chain (herbivores, carnivores, and decomposers).

In like fashion, the end product or end-state of a mechanism may provide start or set up conditions for a subsequent mechanism. Thus single mechanisms are located in wider hierarchical and temporal biological contexts. As we will see, the effort to locate a mechanism within its hierarchical and temporal contexts plays various roles in its discovery. This relationship is especially important for assessing the fit or coherence between the mechanism under consideration and other aspects of the matrix of biological knowledge. Does the mechanism fit with energetic constraints? With what is known about embryological development? With what is known about the evolution of the species? Answering these questions is part of looking to a wider temporal and causal context within which the mechanism came to be or operates.

MULTILEVEL, MULTIFIELD PERSPECTIVE ON MECHANISMS

It is central to our view of mechanistic biology that mechanisms frequently span multiple levels, that adequate descriptions of those mechanisms for many purposes must refer to phenomena at higher levels of mechanisms, that a central goal in mechanistic science is to integrate levels of mechanisms, and that phenomena at higher levels of organization are frequently explanatorily relevant in biology.

It is not part of our view that all explanations must bottom out in some privileged set of fundamental entities and activities (such as elementary particles and strings). Biological explanations rarely need to descend to the depths of quantum physics. As currently understood, most biological mechanisms are blind or otherwise insensitive to differences in particular details of the components at such very small size scales.

Nor is it part of our view that all mechanistic explanations must go down in a hierarchy of mechanisms to be explanatory. In order to explain why a given strand of DNA is on a particular island in the Galapagos, one would do better to investigate the migratory patterns of birds and the selective effects of a recent drought, rather than the molecular components of its DNA. One must, in such cases, ascend from the part to the whole to understand the relevant mechanisms. If one wants to understand why a cyclist's cellular glucose metabolism suddenly accelerated, the right answer might well be that the cyclist has started the climb.

CONCLUSION

In this chapter we characterize mechanisms that produce, underlie, or maintain phenomena. Descriptions of mechanisms specify the mechanism's entities, activities, as well as their temporal, spatial, and active organization. Various features of mechanisms include their productive continuity, regularity, and their location within a hierarchical and temporal context. This perspective stresses multilevel and multifield integration.

We seek mechanisms because knowledge of mechanisms serves the primary aims of science: prediction, explanation, and control. When we know how things work, how the component entities and activities are organized productively such that they underlie the phenomenon to be explained, we can often predict how they will behave. We can answer questions about how they would behave under a variety of different circumstances. We can imagine ways to control their behavior. It is little wonder then that the search for mechanisms has come to dominate the biological sciences generally. The search for mechanisms is tied fundamentally in biology to our ability to know our future, to understand the living world, and to change that world in accordance with our interests.

BIBLIOGRAPHIC DISCUSSION

The beginnings of this mechanistic approach to philosophy of biology trace to the University of Chicago in the 1970s, when Stuart Kauffman (1971) and Bill Wimsatt (1972) analyzed reduction in terms of decomposition into parts. Wimsatt said: "At least in biology, most scientists see their work as explaining types of phenomena by discovering mechanisms, rather than explaining theories by deriving them from or reducing them to other theories, and this is seen by them as reduction, or as integrally tied to it" (Wimsatt 1972, 67). Subsequent University of Chicago graduate students pursued this decompositional view of mechanisms. Bill Bechtel and Bob Richardson wrote the seminal book on heuristics in mechanistic research programs and ways those heuristics can fail when applied to complex systems: *Discovering Complexity: Decomposition and Localization as Strategies in Scientific Research* (1993; 2010). Stuart Glennan (1992; 1996) employed the decompositional analysis of mechanisms to explicate causation: if A causes B, look for the underlying mechanism connecting A and B.

In 1997, independently, two other University of Chicago graduates, Peter Machamer and Lindley Darden, began their own discussion of mechanisms. Darden was working on anomalies for the central dogma of molecular biology and was finding molecular biologists proposing mechanisms for the purported phenomenon of adaptive mutation (Darden 2006, ch. 11). She asked Machamer, an expert on the mechanical philosophy of the seventeenth century, about mechanisms. Carl Craver, Machamer's student writing a dissertation on mechanisms in neuroscience, joined the conversations at an early stage. Our collaboration resulted in the publication of "Thinking About Mechanisms" (MDC 2000). We devised the characterization of mechanisms (used in this chapter) to capture how molecular biologists and neuroscientists use the term. MDC also suggested applying the mechanistic perspective to general issues in philosophy of science on explanation, causation, reduction, and integration.

The MDC paper brought together the two strands of work on mechanisms in philosophy of biology. Darden (2006) suggested that this view should be called the "Chicago School" approach, but that name didn't catch on. Rob Skipper and Roberta Millstein (2005) dubbed it the "new mechanistic philosophy." Jim Tabery (2004) proposed reconciliation between Glennan's analysis in terms of interactions and MDC's approach using activities.

The new mechanistic philosophy is now a lively subarea within philosophy of science. Of course, the topic of causal mechanistic explanation has a longer history. A prominent proponent of mechanistic thinking in science was Wesley Salmon, who explored this idea in his 1984 *Scientific Explanation and the Causal Structure of the World*. Salmon embraces an ontic conception of explanation and attempts to explicate the concept of causation in terms of processes and their interactions. For Salmon, mechanisms are simply processes and interactions (personal communication to LD, 1997). He primarily discussed physics and the "small number of fundamental causal mechanisms, and some extremely comprehensive laws that govern them" (Salmon 1984, 276). Philosophers of biology argue that biology has few, if any, laws (when the term is used to refer to necessary, exceptionless generalizations that apply throughout the universe); see Beatty (1995).

All the original participants continue to further develop the new mechanistic philosophy. Machamer (2004) explicates the concept of activities. Craver (2007) focuses on interlevel integrative mechanistic explanation, which he ties to Salmon's ontic conception of explanation and Woodward's (2003) manipulation view of causation. Darden (2006, ch. 12) pursues strategies for construction, evaluation, and revision of mechanism sketches and schemas. Bechtel and Abrahamsen (2005) contrast the old deductive-nomological (DN) model of explanation with the new mechanistic one. Bechtel extends his analysis to discovery of cell mechanisms (Bechtel 2006) and to analyzing mental mechanisms (Bechtel 2008). Bechtel and Abrahamsen (2010) extend their work to include complex dynamic systems with cycles and feedback loops. Glennan continues to pursue relations between causation and mechanisms, arguing, for example, for a distinction between causal productivity (via activities) and causal relevance (Glennan 2009).

Philosophers debate the extent to which biological mechanisms are regular. Bogen (2005; 2008a) argues for the distinctness of productivity and regularity, claiming that mechanisms need not act regularly. Andersen (2011a; 2011b) argues for various kinds of regularity in mechanisms. DesAutels (2011) assesses problems with regularist and irregularist characterizations of mechanisms and suggests a stochastic middle ground.

Philosophers of biology note that biologists often characterize natural selection as a mechanism, but debate about which concept of mechanism aptly captures that usage. Skipper and Millstein (2005) claim that neither MDC's (2000) nor Glennan's (2002) characterizations of mechanisms adequately explicate the claim that natural selection is a mechanism. Barros (2008) argues against their view, providing an MDC–type account of natural selection as a two level (organism and population) mechanism. A perhaps more fruitful topic is the evolution of mechanisms themselves (see, e.g., Calcott 2009).

In science education, Rosemary Russ and her collaborators (Russ et al. 2008; 2009) applied the MDC view of mechanisms to analyze students' problem-solving episodes. They found that the most sophisticated understanding of the mechanism was indicated, not by ideas about the entities or activities, but by the ability to chain forward or backward. To foster this form of scientific reasoning in the students, Russ et al. (2009) argue, teachers should encourage their students to engage in plausible mechanistic reasoning, even when it deviates from the canonical textbook account at the time.

Thagard (1998; 1999) discusses the search for mechanisms in medicine. Philosophers of medicine debate whether disease mechanisms are to be analyzed separately from normal mechanisms (Nervi 2010; Moghaddam-Taaheri 2011). Another topic in philosophy of medicine is the value of knowledge about mechanisms relative to other kinds of evidence, such as randomized controlled experiments and correlational data (Howick et al. 2010; Howick 2011).

Others extend the new mechanistic philosophy beyond biology. Bill Bechtel (2008) and Paul Thagard (2006; 2012) pioneered the discussion of psychological mechanisms. Hank Frankel's recent books on the history of continental drift (Frankel 2012) are an excellent source of debates about plate tectonics. Frankel says: "The fact that plate tectonics was accepted without a mechanism is one of the great ironies of the overall controversy." (Personal communication to LD.) The issue is discussed in his fourth volume. Jeff Ramsey (2008) works on mechanisms in chemistry. Peter Hedström analyzes and advocates mechanistic explanation in sociology and other social sciences (Hedström and Swedberg 1998; Hedström 2005; Hedström and Ylikoski 2010).

On the issue of aggregativity, discussed in this chapter, see Wimsatt (1997; 2007), and on levels, see Craver (2007, ch. 5, "A Field Guide to Levels"). On functions, see Cummins (1975) and Craver (2001). On working entities in mechanisms, see Darden (2005).

3 REPRESENTING BIOLOGICAL MECHANISMS

INTRODUCTION: REPRESENTING MECHANISMS

Biologists describe mechanisms in many ways. Sometimes they describe mechanisms in narratives that introduce the main characters and describe their activities and interactions as the plot unfolds from beginning to end. Sometimes they depict mechanisms in cartoons and diagrams, where visual conventions (such as boxes and arrows) show how a mechanism works from beginning to end. Sometimes they animate mechanisms in videos showing how the parts move and how the temporal stages unfold one after the other. Sometimes they represent mechanisms in graphical models of causal relations or with equations that show how different magnitudes relevant to the behavior of a mechanism are related to one another. Other times they simulate mechanisms on computers, implement models in robots, or construct scale models of mechanisms in the laboratory.

We refer to representations of mechanisms, whatever form they take, as mechanism schemas (or sometimes mechanistic models). Biologists use mechanism schemas to describe, explain, explore, organize, predict, and control phenomena. These different forms of representation serve different purposes. A movie might be especially good at displaying the temporal organization of a mechanism; a mathematical model might deliver precise predictions; a simulation might help to reveal unanticipated predictions of the model. How scientists represent a mechanism influences how they conceive the problem of discovering it and how they set out to solve that discovery problem. It is therefore useful to dwell a bit on descriptions of mechanisms and on the strengths and weaknesses of different conventions for representing them.

DIMENSIONS OF MECHANISM SCHEMAS

Mechanism schemas vary from one another along four independent dimensions:

- Completeness: Sketch to Schema
- Detail: Abstract to Specific
- Support: How-possibly to How-actually
- Scope: Narrow to Wide

In the following sections we discuss these independent dimensions with the help of a few classic examples from the history of biology.

COMPLETENESS: BLACK BOX SKETCH TO
GRAY BOX SKETCH TO GLASS BOX SCHEMA

Representations of mechanisms vary from one another on a continuum between a *mechanism sketch* and a schema that is *complete enough* for the purposes at hand. A *mechanism schema* is a description of a mechanism, the entities, activities, and organizational features of which are known with sufficient detail that the placeholders in the schema can be filled in as needed.

The central dogma of molecular biology provides a simple illustration:

DNA→RNA→protein

This schema represents the mechanism of protein synthesis. If necessary, one can fill in the details of a specific sequence of bases in DNA, its complementary sequence of bases in RNA, and the particular linear sequence of amino acids in the protein produced by this mechanism. Today, for many instances of protein synthesis, the components of this schema are glass boxes; that is, one can look inside and see the relevant components. The ability to open these black boxes makes the schema especially useful, both for practical applications, such as the creation of genetically modified agricultural strains and the design of gene therapies, and for experimental purposes, such as the creation of genetic knockouts and the investigation of human genetic diversity.

In contrast, a *mechanism sketch* is an incomplete representation of a mechanism. It characterizes some parts, activities, and features of the mechanism's organization, but it has gaps. Sometimes gaps are marked in visual diagrams by black boxes or question marks. Other times, gaps are masked by *filler terms*. Terms such as activate, cause, encode, inhibit, produce, process, and represent are often used to indicate a kind of activity in a mechanism without providing any detail about how that activity is carried out. Black boxes, question marks, and acknowledged filler terms are fruitful because they stand as placeholders indicating where investigators might most productively focus their efforts.

One crucial task in the discovery process is to transform black boxes (for which neither components nor functions are known) into *gray boxes* (for which component functions are specified) into glass boxes. A fully articulated schema consists of glass boxes—they are *complete enough* for the purposes at hand. Such models include all of the entities, properties, activities, and organizational features that are relevant to the pragmatic ends for which the description is being used. Few if any mechanistic models describe all of the entities, activities, and organizational features of a given mechanism. Such descriptions would include so many potential factors that they would be unwieldy for purposes such as prediction and control, and would be unilluminating to limited cognitive agents such as ourselves. Abstraction is often required to bring the intelligible workings of a mechanism into view.

Scientists trying to discover a mechanism are usually quite adept at sketching the target mechanism. Back-of-the-napkin drawings easily convey the places where black and gray boxes reside, some where current work is focused, and others where future work needs to be done. Science teachers, on the other hand, usually work with textbook schemas whose glass boxes have been filled by previous research. An important task in science education is to help students learn to interpret the representations of mechanisms in textbook schemas so that they can recognize the unsolved problems and still puzzling phenomena at the cutting edge of research and acquire the research skills to tackle those problems.

Whether such a drawing counts as a sketch or a schema depends ultimately on whether the represented black boxes and question marks can be replaced at a given stage in the historical development of a model with a more detailed description of the underlying mechanisms. In 1952 James Watson put a piece of paper above his desk with this sketch of the protein synthesis mechanism:

DNA→RNA→protein

This was just before Watson and Francis Crick discovered the double helix structure of DNA in 1953. Nothing was known about the structures and types of RNA and their various roles in the protein synthesis mechanism. In fact biochemists at the time would not have unanimously agreed even that DNA and RNA belonged in a sketch of the protein synthesis mechanism. By 1965, when Watson wrote the textbook, *Molecular Biology of the Gene*, that same drawing was an abstract schema that could be filled with more detail about the structures of the molecules and the activities represented by the arrows: transcription from DNA to messenger RNA and translation from messenger RNA to the order of amino acids in a protein. Subsequent editions of his textbook added more and more details for specifying the initial sketch.

DETAIL: ABSTRACT TO SPECIFIC

Abstraction is the process of dropping details; *specification* is the process of adding them. Contrast Watson's very abstract schema

DNA→RNA→protein

with much more specified diagram of protein synthesis in Figure 3.1.

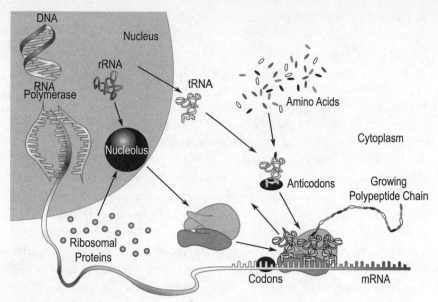

Figure 3.1 Mechanism of protein synthesis.

Clearly the second contains many more details about precisely how DNA makes RNA and how RNA makes proteins. One can envision a continuum of more and more specific representations of the protein synthesis mechanism between these two. At the limit of specification is an *instantiation* of an instance of protein synthesis (not shown), which would specify the DNA base sequence, the complementary base sequence of the messenger RNA, the type of RNA polymerase used to produce mRNA, and the specific amino acids in the growing polypeptide chain of the protein. Abstraction comes in degrees. One and the same mechanism schema or sketch can be represented at different degrees of abstraction for different purposes.

Assume that one has a schema composed of glass boxes that can be opened as necessary to reveal the mechanism's productive continuity. The degree of detail depicted in a particular representation of a mechanism varies depending on one's practical needs. Differing amounts of detail are appropriate for instructing elementary school students, writing a college textbook, delivering a conference presentation, designing an experiment, building a piece of equipment, and preparing for brain surgery. For instructional purposes, it is often useful to abstract away from all of the messy details to convey the more abstract features and the overall organization of the mechanism's workings. For some investigative purposes, one needs to present a very detailed description of the mechanism under investigation in order communicate to others how to reliably produce or detect some effect of interest. While we cannot specify in general what will count as complete enough across these different contexts, it is nonetheless possible to say that the schema should include everything that makes a significant difference for the purposes at hand. Added detail may allow one to increase significantly the probability of completing a surgical procedure without doing damage to neighboring brain structures. More detail might be required to show how to intervene to change the value of one variable without changing the value of any other so as to detect that lone variable's contribution to a mechanism. But such detail is superfluous when it introduces minute causal factors (whose impact on the mechanism is too small to be of consequence), improbable causal factors (whose impact is too unlikely to be of consequence), or atypical causal factors (whose impact is idiosyncratic to particular organisms and not representative of the group under investigation). So the question is this: Does the presence or absence of the detail in question change the predicted behavior of the mechanism in a way that (or to an extent that) matters for one's purposes? If the answer is yes, then it should be included in a description that is complete enough for those purposes. If not, then such detail is effectively screened-off; the detail is irrelevant.

SUPPORT: HOW-POSSIBLY, HOW-PLAUSIBLY, AND HOW-ACTUALLY

Mechanism schemas also vary in their degree of evidential support and accuracy. A well-supported schema satisfies known constraints on components, activities, and their organization with one another, and it coheres with other well-supported, non-rival theories. How-possibly schemas are loosely constrained conjectures about how the mechanism works. They describe how a set of parts and activities might be organized such that they exhibit the explanandum phenomenon. It may be unknown whether the conjectured parts exist and, if they do, whether they can engage in the activities attributed to them in the schema. Some computational models, for example, are purely how-possibly schemas. For example, one might simulate human motion detection in a computer program written in the computer programming language LISP. Such a simulation might embody no commitment to the idea that the brain is somehow executing the basic operations of LISP (such as CARs and CDRs). How-possibly schemas are often heuristically useful in directing research programs in the design of experiments.

However, if one's goal is to control, explain, and/or predict how a mechanism will behave under the widest variety of conditions, one requires more than a mere how-possibly schema. For those purposes it is often crucial to know how the mechanism really works. One wants a how-actually schema (or at least a schema that satisfies most or all of the known constraints on the mechanism, or that identifies most or all of the features of the mechanism that make a difference big or probable enough to be of relevance to the problem at hand). *How-actually schemas* describe real components, activities, and organizational features of the mechanism.

Between how-possibly and how-actually lies a range of *how-plausibly* schemas that are more or less consistent with more or fewer known constraints on the components of the mechanism. As the number and variety of constraints increases, more and more possible schemas become implausible. As more empirical findings are credentialed and applied, more and more possible schemas become implausible. As additional evaluative strategies are utilized (such as testing for coherence with fit within the context of other schemas and theories), more and more possible schemas become implausible. In complex cases, different schemas might have counterbalancing strengths and weaknesses: each model is superior to the other with respect to some subset of the constraints.

We hasten to emphasize that mechanism schemas frequently contain idealized (that is, false) assumptions about the entities, activities, and organizational features of the mechanism. One might assume, for example, that the concentration of an ion in a cell is everywhere uniform, or that the axon of a cell is shaped like a cylinder, or that organisms in a population mate perfectly randomly. In such cases, the scientist has to make a decision about whether the idealizing assumptions matter (that is, make a difference) for the purpose to which the schema is to be put. A literally false schema, that is, might be plausible enough for a given purpose even if it is not, technically, a how-actually schema. Scientists often stop short of developing how-actually schemas and are satisfied instead with identifying a region of the space of possible mechanisms within which the actual mechanism falls. Idealizing models often help to identify that region of the space of possibilities, the space of possibilities that are how-actually-enough.

SCOPE: NARROW TO WIDE

Finally, mechanism schemas vary from one another in their scope, namely the size of the domain to which the schema applies. A mechanism schema may represent a single unique case (e.g., a mechanism producing a unique historical event, such as the origin of the AIDS virus or the extinction of the dinosaurs), or a recurring mechanism in only one species or in a type of cell or tissue (e.g., a mechanism for small cell carcinoma in the human lung). More often, mechanism schemas in biology have a middle range of applicability, namely, they apply to some subset of biological cases, such as memory mechanisms in the hippocampus of vertebrates, or to the set of genes in some species of bacteria that metabolize milk.

Only a few biological mechanisms are found in all living things on earth. Watson and Crick's three-place schema for protein synthesis (the central dogma schema discussed above) is one of the most highly conserved mechanisms in the living world. The scope of the schema includes almost all instances of protein synthesis in living organisms on earth (some viruses are exceptions). Specifications of this schema can be used to represent variations in the protein synthesis mechanisms in more specific groups. Bacteria, for example, use different enzymes and have smaller ribosomal RNA than other organisms. Thus, representations of mechanisms can be applied to domains of more restricted range as the details are specified and finally, in the limit, can be instantiated for one specific case. Conversely, beginning with a single case, one can abstract away, dropping details and test to see if the more abstract schema has a wider scope of applicability.

Scope and abstraction often vary with one another. Increasing the degree of abstraction of a how-actually schema often produces a schema with a higher degree of generality because dropping details allows additional kinds of cases to be represented by the schema. But degree of abstraction and degree of generality are independent dimensions along which schemas vary. The schema

DNA→RNA→protein

applies in nearly every organism. The schema

RNA→DNA→RNA→protein

is instantiated only in the domain of retroviruses. The schema

DNA→protein

(in which DNA produces proteins directly without RNA as an intermediate) is so far as we know not found outside the laboratory.

Arguably the mechanism of natural selection applies anywhere in the universe and at all times. Hypothetically, if alien life forms reproduce and do so in a way that produces variants with different fitness values in their environments, then the mechanism of natural selection would be expected to operate.

When mechanisms of very wide scope are found, biologists generate hypotheses to account for the lack of variability. One example is the genetic code, which consists of the relations between three bases of DNA and one amino acid that is utilized during protein synthesis during the translation of messenger RNA. One hypothesis about why the genetic code is found so widely is that, after some period of development when certain features might have been adaptive, the code became frozen. Once things began working together one way (in a common ancestor to all living things), then changing them was too costly. Other hypotheses look for adaptive or physiochemical constraints to explain the origin of this particular code and its maintenance as a nearly universal code. In general, when less variability than expected is found, a hypothesis is that selection has operated to maintain an adaptive, working mechanism in the face of constantly occurring mutations that otherwise would produce more variability.

Schemas for modules within mechanisms (that is, complex subcomponents) sometimes have wider scope than do the schemas for the mechanisms of which they are components. This is because evolution often reuses old modules for new purposes. Many of the most widely conserved genes are components of the translation module in the protein synthesis mechanism, while some genes play roles in DNA replication. The specific components of these wide scope mechanisms are not the same, but many modules have been conserved during evolution. Often, similar sequences in two genes indicate that they have the same functional role in the same kind of mechanism. A surprising finding was that a protein (called RHO) used by fruit flies for cell signaling has a similar structure to a protein in bacteria used for signaling between different bacteria; in fact, they are sufficiently similar that the protein from the fruit fly, when added to mutated bacteria, can restore the bacteria's ability to signal. (Bacteria sense other bacteria in their environment and use those signals for various functions.) However, sometimes evolution co-opts a gene and uses it for a different function. Proteins in the crystalline lens of the eye originated as metabolic enzymes, with a completely different function. Hence, similar sequences provide a good first hypothesis that the genes play similar roles in similar mechanisms, but further empirical work might show that there is a change in the mechanism context in which the module functions. Modules are copied, edited, and moved during evolutionary change. For example, a module with a similar DNA sequence might have a conserved function when it is moved into a new context.

Issues about the scope of schemas and modules are especially important in experimental research. The issue arises when one makes inferences from model organisms and model experimental systems to a target mechanism. In medical research, which seeks the best model for a target human population, this question is of considerable practical significance. Different concerns arise when the goal is fundamental biological research to find a model system that will generalize to a schema of a wide domain, such as that of protein synthesis.

In sum, mechanism schemas vary from one another independently in their completeness, their degree of abstraction, their support, and their scope. These dimensions apply equally to all of the different formats (computational, mathematical, pictorial, and verbal) in which a given mechanism might be represented. We close this chapter by considering some of the conventions and norms used to build visual and mathematical descriptions of mechanisms.

VISUAL REPRESENTATIONS OF MECHANISMS

Mechanisms are routinely represented visually in diagrams showing how the mechanism works from beginning to end. There are many strategies for producing visual representations of mechanisms. Often single diagrams employ many representational methods at once. We list a few such conventions.

- *Abstract from Detail*
  Visual representations of mechanisms typically abstract from the gory details of a particular mechanism. Background conditions are often omitted to focus on just the key players. Often a single icon represents a population of entities (such as neurotransmitters or receptors). One activity is shown where many are operating, or activities are omitted or represented by arrows. Such abstraction makes a mechanism intelligible by representing it in a form that can readily be analyzed by the human visual system.
- *Represent Entities and Activities*
  Visual diagrams of mechanisms represent the key players, the entities and activities, by which the mechanism works. Entities are often represented iconically, with more or less attention to the particular structural features of the entities in question (a molecule, for example, might be represented by a blob or box). In general, it is easier to draw entities than to draw the activities in which they engage; comparably few conventions have been established to represent activities. Sometimes arrows are used generically to depict the operation of an activity. Some fields develop standardized icons for specific kinds of activities. In chemistry, solid lines represent covalent bonds, while dashed lines represent weaker hydrogen bonds.
- *Represent the Temporal Order*
  Diagrams often show paragraphs of activity that render the working of the mechanism intelligible. In some diagrams, separate frames of a cartoon represent sequential stages or extended events in the working of a mechanism. In other diagrams, artists use spatial relations to represent temporal sequences: e.g., the temporal stages of a mechanism unfold from left to right; or sequential stages (boxes) connected by arrows exhibit the temporal order. Parallel pathways are also represented spatially with branches and joins, while cycles are represented with arrows that loop back to early stages of the mechanism. In each case, the success of the diagram turns

in part on representing intelligible paragraphs of activity that the viewer can see as meaningful stages in the mechanism.

- *Show Upper and Lower Levels and Their Relations*
  Often it is helpful to represent multiple levels of mechanism in a single diagram. The most straightforward convention is to use spatial compartmentalization to represent a move to a lower level: the lower-level item is drawn within the represented boundaries of a higher-level item. In other cases, levels are represented by a telescoping relation in which a higher-level component is visually expanded, as if looking through a microscope, to reveal its internal parts and activities. A call-out box might be used, as on a map, to show the region of one diagram that is exploded in another diagram.

- *Outline Global Organization*
  Diagrams often let one see in one glance how the overall mechanism is organized, showing the spatial relations among the components, the temporal order of the stages, and the active organization, i.e., what acts on and interacts with what. That is, visual diagrams assemble all of the pertinent information into a single snapshot (or a series of snapshots) exhibiting how the whole mechanism works from beginning to end.

Many of these representation techniques are exemplified beautifully in Figure 3.2, which represents part of the mechanism by which a sperm fertilizes an egg.

The picture in the upper right corner is a photomicrograph of many sperm, with their small heads and long tails, swimming around an egg. The photo shows an intermediate stage between the egg attracting sperm and the moment a single sperm penetrates the cytoplasm of the egg. It shows *the structure of entities* in the mechanism and their *relative sizes*: for example, the many small sperm and the large egg. The photograph also contains detail that is not obviously relevant to the mechanism at hand. It shows, for example, the granulose cells surrounding the egg, the textured invaginations of the egg surface, and sperm scattered hither and yon. Drawings abstract away from obfuscating detail to focus attention only on the *working entities*, their crucial features, and the relevant organizational features. As a snapshot in time, without annotation, a photograph cannot show a mechanism's *temporal stages*. Because of these limitations, the photograph is much less informative about the mechanism than the two diagrams that occupy most of the figure.

Focus therefore on the drawing. The small circle at the top left depicts a cross-section of the egg, with several sperm surrounding its outermost layer. This illustration connects the photomicrograph with the detailed representa-

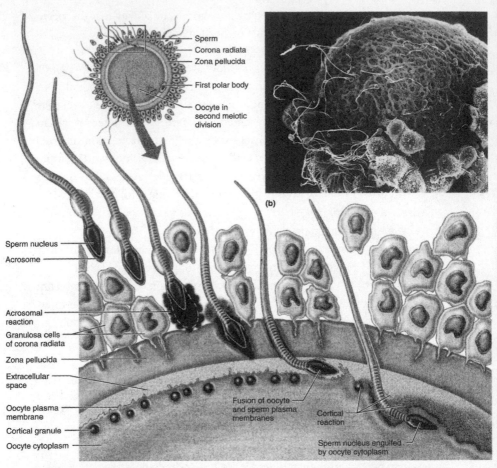

Figure 3.2 Multiple visual representations of a sperm penetrating an egg. (Marieb 2004) Reprinted by permission of Pearson Education, Inc., Upper Saddle River, NJ.

tion below. The rectangular call-out area at the edge of the egg is enlarged in the lower diagram to show the mechanism by which a sperm penetrates an egg. The arrow between the cross-section of the egg and the larger diagram of the penetration stages directs attention to a *lower level*. The lower diagram abstracts from irrelevant details to show just the key stages in the mechanism by which a single sperm penetrates the egg. The representation of the mechanism starts after an earlier stage, during which the egg secretes a chemical to attract sperm of only the appropriate species. (This is one of the many isolating mechanisms that keep species distinct.) Of course, myriad sperm arrived earlier, but the unsuccessful sperm don't make it into the drawing. Nor does the drawing show the enzymes that the sperm use to break down the outer layer of the egg's membrane and expose it to penetration by a sperm.

The diagram effectively depicts some of the *temporal stages* as the sperm approaches and ultimately invades the egg's cytoplasm. Each of the six drawings represents a *paragraph of activity*, ordered from left to right to follow the sperm's progress through time and space. A series of granulosa cells of the egg's outer layer are shown moving out of the way as the sperm wiggles into place. The third drawing shows an advancing cloud in front of the sperm, which represents the release of enzymes that dissolve both the outer layers of the sperm and the next layer of the egg's membrane. This *lower-level activity* of enzymatic lysis, as it is called, is only suggestively represented with the advancing cloud in the fourth stage. The fifth drawing illustrates the fusion of the egg and sperm membranes. The last drawing has two parts. First, the small circle and the chemicals moving from it into the space above the last sperm head represent the reaction that blocks other sperm from entering the egg. Second, the sperm head is shown entering the egg cytoplasm, on its way to later stages, not shown, where its chromosomes are released and fuse with those of the egg to form the zygote and complete the mechanism of fertilization. The *overall organization of the mechanism* of penetration, from the sperm approaching the egg, to its entry into the egg cytoplasm, is clearly shown from beginning to end.

Pictorial diagrams exploit the close connection between understanding a mechanism and being able to visualize how it works. Skillful scientific artists exploit features of the visual system to efficiently convey information about how a mechanism works from beginning to end. Notice, however, that this diagram does not contain information about quantities, magnitudes, rates, and durations. It does not tell us how many sperm approach the egg, or the concentration of the enzymes required to break down the membrane, or how the concentration of enzymes affects the probability of penetration. Such quantitative features of a mechanism often must be represented with mathematical conventions, to which we now turn our attention.

MATHEMATICAL AND GRAPHICAL REPRESENTATIONS OF MECHANISMS

In mathematical descriptions of mechanisms, key features of a mechanism are represented as variables or constants. Mathematical relations among the variables are then used to capture how magnitudes (i.e., the measurable quantities) in the mechanism change in relation to one another. A typical convention is to represent independent variables (often standing for causes) on the right-hand side of an equation and dependent variables (often standing for effects) on the left-hand side. Mathematical models might be used to describe precisely the phenomenon for which a mechanism is to be sought (as phenotypic ratios describe a hereditary phenomenon for which a genetic mechanism might be sought). Other times they are used to characterize the components in a mechanism and their interactions with one another.

Below, we discuss Hodgkin and Huxley's mathematical model of the dynamic activities involved in the generation of action potentials. However, mathematical models can be found in many areas of biology. Epidemiologists use mathematical models to characterize how a disease spreads through a population. Evolutionary biologists use mathematical models to simulate population dynamics under a wide variety of conditions. Kidney physiologists use mathematical models to characterize how the blood is filtered under varying hormonal and ionic concentrations. Ecologists use mathematical models to predict the effect of deforestation in rain forests. In many cases, such mathematical models are used to characterize mechanisms and the phenomena they explain.

Consider some of the useful features of mathematical representations of mechanisms:

- *Exhibiting Quantitative Relations Among Variables*
  Equations are quantitative descriptions of how different variables in a system are related. Pictorial diagrams often do not capture the precise interrelations of magnitudes that are often crucial to understanding how a mechanism works.
- *Abstracting from Details*
  Like pictorial representations, mathematical equations and graphs abstract away from many details of the entities and activities in a mechanism and aspects of their organization in order to focus on a few key magnitudes. However, sometimes they capture some aspects of the temporal order and active organization of the mechanism that are less easily represented in visual diagrams.
- *Making Assumptions Explicit*
  Mathematical representations are formal, and scientists must make explicit how the components in the mathematical model relate to features of the mechanism. One must chose, for example, which features of the mechanisms to represent in variables. One must choose whether the variables are discrete or continuous. To generate elegant and practically useful equations, one must often make idealizing assumptions. The task of building a mathematical model forces one to be painfully explicit about these basic assumptions.

- *Demonstrating Sufficiency*

  Mathematical representations of mechanisms are often used as tools for demonstrating that a set of components, organized as the schema suggests, would indeed suffice to produce the phenomenon. One can demonstrate, for example, that the equations descriptive of the parts of the mechanism and their activities would or would not, under reasonable assumptions about the values of the relevant variables, entail a mathematical description of the phenomenon, that is, the behavior of the mechanism as a whole. Verbal descriptions and pictorial representations often lack this demonstrative character.

- *Dealing with Complexity*

  While visual diagrams are useful in part because they allow people to literally see how a mechanism works from beginning to end, the human visual system is easily overwhelmed as mechanisms become more complex. As the number of components and activities increases, or as components have multiple effects on one another, or as components engage in dynamic interactions with one another, one can no longer see the mechanism working because the visual system cannot keep track of all the relevant changes within and among the components of the mechanisms. Mathematical models can include arbitrarily many variables and arbitrarily many relations among them. They can also be used to represent dynamic forms of change the human visual system is not well equipped to model.

- *Making Precise Predictions*

  Mathematical representations often give one the ability to make precise predictions about quantitative features of a mechanism and so help in testing the representation rigorously against experimental results. Mathematical models can be used to generate predictions in particular about how the mechanism would behave if it were placed in conditions that have not yet been observed.

- *Allow What-if Modeling*

  Mathematical models may allow exploration of what-if scenarios that cannot be instantiated in experimental systems, either because such set-ups cannot be physically constructed or because to do so would be too costly in time and resources. Mathematical models are often a first step in building computational simulations of a system.

The Hodgkin-Huxley model of the action potential is a particularly rich and compelling example of a mathematical description of a mechanism. Action potentials are electrical signals in neurons. More specifically, they are fleeting changes in *voltage* across a neuronal membrane, as represented in the solid curve in Figure 3.3. Voltage can be understood metaphorically as a "tension" or a "pressure" that can move *charges* from one side of the membrane to another. The charges in this case are borne by positively charged particles, known as *ions*. The movement of ions across the membrane constitutes the flow of an electrical *current*. Alan L. Hodgkin (1914–1998) and Andrew F. Huxley (1917–2012) argued that action potentials are produced through the coordinated movement of ions (that is, ionic currents) across the membrane. The Hodgkin and Huxley model describes the currents that make up the action potential.

In its rest state the membrane of a neuron is negatively charged relative to its extracellular environment. This negative resting state of the membrane (we now know) is produced and maintained by an active molecular pump that exchanges three positively charged sodium (Na^+) ions from inside the cell for two positively charged potassium (K^+) ions from outside the cell. With each such exchange, the inside of the cell becomes one unit of charge more negative with respect to the outside of the membrane. The pump thus sets up a voltage (or potential) difference between the two that can be used to drive charges (that is, currents) from one side of the membrane to the other. This resting membrane potential (or resting voltage) is the set-up condition for the generation of action potentials; this is why the membrane voltage begins at around −50 mV (rather than at 0 mV) in Figure 3.3.

The solid line in Figure 3.3 represents the action potential phenomenon, the change in voltage across the membrane over time. During an action potential, the voltage across the membrane first rises steeply from −50 mV until it peaks around +40 mV. Then it reverses, dropping precipitously until it eventually becomes more negative than it was at rest. In the final stage, the cell gradually returns to its resting voltage. The shape of this curve can be characterized precisely in terms of, for example, the slope of the rising and falling phases, the peak voltage, the trough voltage, and the overall time course of the observed changes in membrane voltage.

Over many years spanning the Second World War, Hodgkin and Huxley gradually pieced together a schema for the mechanism of the action potential. According to that schema, action potentials are produced by local and fleeting changes in the membrane's permeability to the flow of sodium and potassium ions. In electrical terms, they are produced by local increases in the membrane's *conductance* to the flow of these ionic currents over time or, reciprocally, local reductions in the membrane's *resistance* to the flow of such currents over time. Hodgkin and Huxley's mathematical model shows how voltage, conductance/resistance, and current flow change over time during the course of an action potential.

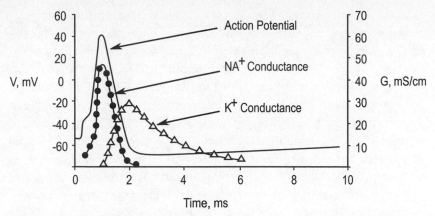

Figure 3.3 The action potential. The solid line represents change in membrane voltage (V) over time. Voltage is measured in millivolts (mV), as shown on the left. The curve plotted over circles represents changes in Na+ conductance over time. The curve plotted over triangles represents changes in K+ conductance over time. Conductance is measured in millisiemens (mS), as shown on the right.

For now, however, let's continue with a more qualitative characterization of their schema. An action potential is triggered when the membrane voltage rises beyond a threshold value as a result of electrical input into the cell (from other cells or from an experimenter's electrode). Beyond a certain threshold in membrane voltage, the membrane decreases its resistance (or, reciprocally, increases its conductance) to the flow of sodium ions across the membrane. This increase in sodium conductance is drawn in the curve plotted over circles in Figure 3.3. As the conductance of the membrane to sodium increases, sodium ions then begin to flow into the cell. Two activities are involved in this movement. First, sodium *diffuses* into the cell because the concentration of sodium outside of the cell is much greater than the concentration inside. Second, sodium is *electrostatically attracted* into the cell because sodium ions are positively charged and the interior of the cell, when it is below 0 mV, is negative (there is thus an electrical force on sodium). The sum of these two forces is known as the *driving force*. This movement of charged particles across the membrane constitutes an inward current, the sodium current. A consequence of this movement is that the voltage across the membrane becomes less negative, and eventually more positive. (Note that the curve for the action potential goes above 0 mV).

When the force of diffusion and the electrostatic force are balanced, the membrane is at equilibrium and there will be no net flow of sodium (this voltage is known as the sodium *equilibrium potential*). The equilibrium potential for sodium is above the peak of the action potential. The cell would eventually reach the sodium equilibrium potential under these conditions were it not for two things that happen next. The rising membrane voltage causes both. First, the membrane's conductance to sodium begins to inactivate (as shown in the curve fitted to circles in Figure 3.3). Second, the membrane's conductance to potassium begins to increase (as shown in the curve fitted to triangles in Figure 3.3). Potassium diffuses and, indeed, is repelled out of the increasingly positive cell, carrying its positive charges with it. These two processes combine to drive the membrane voltage back down below the resting potential (toward the equilibrium potential for potassium). Eventually, the membrane's conductance to potassium returns to its rest value, and the sodium-for-potassium pump returns the membrane to its resting potential. This is the termination condition for the action potential.

This verbal schema captures many (but not all) of the main players in the action potential mechanism. It provides the background for understanding Hodgkin and Huxley's mathematical model. Their mathematical model combines physical facts about electrical circuits (such as voltage, current, and conductance/resistance) with distinctively biological facts about the characteristic changes in membrane conductance that underlie the action potential.

Hodgkin and Huxley built their model under the ingenious assumption that the membrane can be described as an electrical circuit, composed of batteries, currents, resistances, and the like. Their "equivalent circuit" diagram of the membrane is shown in Figure 3.4.

The centerpiece of Hodgkin and Huxley's mathematical model is known as the Total Current Equation:

$$I = C_M dV / dt + \boxed{G_k n^4 (V - V_K) + G_{Na} m^3 h (V - V_{Na})} + G_1 (V - V_1)$$

This equation is a sum of four currents: left to right, the capacitive current, the potassium current, the sodium current, and the leakage current. For now, we limit our attention to the expressions inside the box. (The first addend [before the box] describes the capacitive current. This equation describes an apparent flow of ionic current that occurs as oppositely charged ions are attracted to one another across the cell membrane. This capacitive current flows until the capacitor [in this case, the membrane] has stored as much charge as it can [the membrane capacitance, C]. The final addend [after the box] describes the leakage current. This is the net current produced by ions other than sodium and potassium.) Because the equations for the potassium and sodium currents are structurally identical, what we say about the first applies, *mutatis mutandis*, to the other. Accordingly, we discuss only the first of these, the potassium current.

This part of the total current equation has three components multiplied

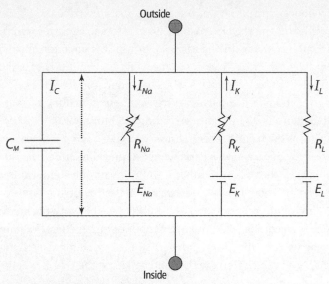

Outside

Inside

Figure 3.4 Hodgkin and Huxley's Equivalent Circuit Diagram of the neuronal membrane. I_C, I_{Na}, I_K, and I_L represent capacitive, sodium, potassium, and leakage currents. C_M represents a capacitor. The Rs represent the resistance (the inverse of conductance) to sodium, potassium, and leakage. The arrows across the resistors indicate that their resistance changes during generation of an action potential. E represents the resting membrane potential (V_{rest} in the text), and E_{Na}, E_K, and E_L represent batteries, which correspond to the equilibrium potentials for sodium, potassium, and leakage.

together: G_K, n^4, and $(V - V_K)$. G_K represents the maximum conductance (or, reciprocally, the minimum resistance, R) of the membrane to potassium. The second factor, n^4, is a summary of experimental results about the distinctively biological behavior of the membrane; it describes precisely how the membrane's conductance/resistance changes as a function of voltage and time. (Hodgkin and Huxley's primary theoretical achievement was to write equations for n, m, and h that accurately and precisely express these experimental results.) The multiplicative product of G_K and n^4 describes in precise mathematical terms how the membrane's conductance to potassium changes with voltage and time (as shown graphically the curve plotted over triangles in Figure 3.3).

The final factor in Hodgkin and Huxley's model of the potassium current, $V - V_K$, corresponds to the battery in the equivalent circuit diagram. It represents, in a sense, the net forces driving the potassium current. (Literally, the equation represents the difference between two differences: V is the difference between the membrane voltage and the membrane voltage at rest, and V_K is the difference between the equilibrium potential for potassium and the membrane voltage at rest.) As described above, this portion of the equation summarizes the effects of two activities: diffusion and the electrostatic driving force. The strengths of these two forces (combined in the driving force) determine the direction and the strength of current flow through a fixed resistance. Of course, in this case the resistance is not fixed; it is changing as described by G_K and n^4.

With these interpretations in hand, it is easy to see the equation for the potassium current as an instance of Ohm's law. According to Ohm's law, $I = V/R$. Because conductance (g) is the reciprocal of resistance ($R = 1/g$), $I_K = g_K V$. Because, as described above, the conductance is variable in biological membranes, this needs to be rewritten as $I_K = G_K n^4 V$. (Remember, G_K is the maximum conductance; n^4 has a peak value of 1.) And because we are focused on the net driving force for potassium, the V term is in fact represented as the difference between the membrane voltage (V) and the equilibrium potential for potassium (which Hodgkin and Huxley represent as V_K). Thus, $I_K = G_K n^4 (V - V_K)$. (The equation describing the sodium current has the same form, with m and h playing the role of n in the potassium equation.)

Putting it all together, the total current equation is a sum of four equations, each describing a different current flowing across the membrane. The variables in these equations home in upon just a few key magnitudes that are especially central to the mechanism. Hodgkin and Huxley then borrow the formal tools of electrical circuit theory (Ohm's law) to generate equations that describe how these variables change as a function of one another in the course of an action potential. The result is a mathematical model that more or less accurately predicts many of the key electrophysiological properties of a neuron, including its peak and trough, its slopes of rise and decline, and a variety of electrophysiological features of nerve cells below the threshold for an action potential.

The example illustrates some of the central features of mathematical models. It uses variables and mathematical relations to represent relevant magnitudes and dynamic relations in the mechanism. It abstracts away from many of the details about the generation of nerve cells to focus on a few key variables. For example, the model leaves unspecified the mechanism by which the membrane changes its conductance. These mechanisms were not known at the time, and Hodgkin and Huxley did not need to know them for their purposes (more about which below). The effort to build this mathematical model forced Hodgkin and Huxley to be explicit about the assumptions of their model: that the key features of the mechanism can be represented as components of electrical circuits, that the appropriate circuit diagram is the one represented in Figure 3.4, that Ohm's law applies to neurons, that the relative concentrations of ions inside and outside the cell have particular values, that the neuron is at a particular temperature.

It was especially important to Hodgkin and Huxley that these variables describing components be filled with values that conform to the results of their experiments. That is, the concentrations determining the driving force on potassium had to match the measured concentrations, and the changes in membrane conductance had to match the results of Hodgkin and Huxley's laborious voltage clamp experiments. They were, after all, attempting to describe the actual mechanism by which neurotransmitters are generated, not merely a how-possibly mechanism. Their fundamental purpose in building this model, in fact, was to show that a membrane mechanism, working as the equations describe and under the conditions present in real-live nerve cells, would change the membrane voltage precisely as shown in the solid curve of Figure 3.3. The model thus serves the demonstrative job of showing that such a mechanism would, under these assumptions, produce action potentials. And given that Hodgkin and Huxley had independent evidence that the values and relations in the model in fact corresponded to the magnitudes and regularities observed in this mechanism, they had good reason to embrace their schema: action potentials appear to be generated by a flux of ions across a membrane as the membrane changes its resistance with voltage and time.

The model also describes something about the mechanism that cannot (easily or efficiently) be described in words or illustrated in drawings. First, it shows precisely how the different key magnitudes in this mechanism change over time, mapping values of one variable onto values of the others. In doing so, it tells us something crucial about the active organization of a mechanism: it tells us how much and in precisely what respects the different variables affect one another. The verbal expression that voltage changes the membrane conductance, in other words, does not contain within it a precise characterization about how unit differences in voltage translate into unit differences in conductance. Second, the model allows one to represent simultaneously the many and promiscuous interactions among these different magnitudes. Part of the reason that this mathematical representation is so useful is that all of the key factors described in Hodgkin and Huxley's model—the membrane voltage, the resistances, the concentrations of ions—are changing at once and as a function of one another. The mathematical representation describes precisely how the relevant features of the different components in this mechanism change one another together at once, and precisely what the net result of all of those interactions will be.

Finally, this mathematical model can be used to generate precise predictions about how the action potential will change under a variety of conditions. One might ask, as electrophysiology professors routinely do on exams, how the action potential would change if one were to double the extracellular sodium concentration, or to block potassium channels with a toxin, or to cut the sodium conductance by a third, or to raise the temperature of the cell by five degrees. The model can be used to answer all of these questions. A number of different simulations of the Hodgkin-Huxley model are widely available that allow one to explore the predictive consequences of their model, pedagogically experimenting with the model to see how parameter settings influence the behavior of the model as a whole. The value of the model as a demonstration of the sufficiency of the schema (discussed above) depends crucially on the fact that the model makes correct predictions about a large range of phenomena.

Mathematical models cannot do everything that can be done with words or in visual diagrams. Qualitative thinking was crucial for Hodgkin and Huxley's discovery. The rise of electrophysiology, the very idea that nerve cells traffic in electricity, was not first and foremost a mathematical hypothesis but a qualitative one. The equivalent circuit diagram should be seen as a translation manual for connecting aspects of that qualitative description with the mathematical representational system of electrical circuit theory. The mathematical theory adds precision and rigor to the qualitative description, capturing features of the active organization and dynamics of the mechanism. The qualitative theories (in diagrams and in words) give content to the variables of the model by connecting those variables to concrete entities, activities, and organizational features of the target mechanism. The Hodgkin and Huxley model, absent such a qualitative interpretation, would be empty. Biology in particular is a domain in which qualitative information about components and their organization can sometimes be combined with abstract representations of the mathematical relations among the magnitudes relevant to how the mechanism works. Qualitative models without mathematical models often lack precision; mathematical representations without qualitative interpretations lack biological content.

CONCLUSION

Scientists represent mechanisms in a variety of formats: in narratives, schemas, diagrams, videos, simulations, graphs, and mathematical models. They produce schemas that can vary considerably from one another in their completeness, detail, support, and scope. Various conventions aid in representing mechanisms visually. Highly simplified and idealized graphical and mathematical representations depict coordinated changes among some of the parts. Mathematical and computational models may mirror more or less accurately how the mechanism actually works, allowing exploration of what-if scenarios. Different representational formats play different roles in the search for mechanisms.

BIBLIOGRAPHIC DISCUSSION

The distinction between a schema and a sketch was introduced in MDC 2000. On methods of forming abstractions and further discussion of the abstraction for natural selection, see Darden and Cain (1989). For more on filler terms, see Craver (2007; 2008a); Kaplan and Craver (2011). We took the terminology of black, gray, and glass boxes from Hanson (1963), who used the terms in a slightly different way. The idea of middle range models in biology is found in Schaffner (1993). Wimsatt (1987) discussed ways that false models can be useful. See Woodward (2003) on difference-making and relevance.

On Watson's and Crick's different representations of the central dogma, see Keyes (1999). Francis Crick (1967) speculated that the genetic code, after a period of adaptive development, might have become frozen, with additional changes too costly. For more recent discussion of the evolution of the genetic code, see Koonin and Novozhilov (2009). Morange (2006) discusses single molecules that behave like little machines. Such molecular machines are typically entities within the kinds of mechanisms discussed here, although such changes in a single molecule might be explained by a mechanism made up of the molecule's components and their interactions in the molecule's chemical environment.

Hodgkin and Huxley published their classic series of experiments in 1952 in volume 117 of the *Journal of Physiology*. The last of these papers, "A Quantitative Description of Membrane Current and Its Application to Conduction and Excitation in Nerve," contains the mathematical model. The earlier papers describe the electrophysiological experiments that made the modeling possible. Autobiographical descriptions of this episode can be found in Hodgkin (1992) and Huxley (1963). For recent philosophical discussions of the Hodgkin-Huxley model and its relevance to the search for mechanisms, see Bogen (2005; 2008b), Craver (2006; 2007; 2008a), Schaffner (2008), and Weber (2008).

For strategies of visual representation in science, see Tufte (1989). Maria Trumpler (1997) offers a beautiful analysis of visual representations in the history of research on the action potential.

4 CHARACTERIZING THE PHENOMENON

A mechanism is always a mechanism *of* a given phenomenon. The mechanism of protein synthesis synthesizes proteins (macromolecules composed of amino acids). The mechanism of the action potential generates action potentials (rapid changes in the electrical potential difference across a neuron's membrane). For this reason, descriptions of the phenomenon constrain the character of the mechanism that one will seek to discover; they describe what the mechanism has to be able to do. Similarly, the boundaries of mechanisms—what is in the mechanism and what is not—are specified by reference to the phenomenon that the mechanism explains. Those entities, activities, and organizational features that are in the mechanism are those whose activities or properties are relevant (in the sense to be developed in later chapters) to the phenomenon to be explained. In a slogan, mechanisms are the mechanisms of the things that they do.

CHARACTERIZATIONS OF PHENOMENA
SHAPE THE DISCOVERY OF MECHANISMS

Characterizing the phenomenon to be explained is a vital step in the discovery of mechanisms. Characterizing the phenomenon prunes the space of possible mechanisms (because the mechanism must explain the phenomenon) and loosely guides the construction of this hypothesis space (because certain phenomena are suggestive of possible mechanisms). To describe a phenomenon is to characterize it in the language of a given field and to implicitly call up the host of explanatory concepts, the store of entities, activities, and organizational structures known to a field at a time, that might be used to construct a schema of the mechanism.

Consider the phenomenon of spatial memory as it is studied by neuroscientists and experimental psychologists. It is not at all obvious in advance of considerable empirical inquiry that there is any such phenomenon—spatial memory—for which a distinct mechanism exists. Given that there is such a faculty or phenomenon, it is not at all obvious in advance of considerable empirical inquiry how that phenomenon is properly to be characterized. Debates over the taxonomy of memory can be understood as debates about how to characterize and individuate different memory phenomena. The character of the phenomenon is open to revision in light of evidence. Episodes of mechanism discovery typically involve at least some effort to recharacterize the phenomenon as scientists learn more and more about its mechanism.

Edward Tolman's (1886–1959) now famous experiments on maze learning, performed in the 1930s and 1940s, are a good example. Tolman's work was instrumental in shaping what contemporary neurobiologists think about spatial memory. Rats trained to successfully navigate a contorted and circuitous route through a maze (Figure 4.1a) were subsequently placed into the same maze with a new, more direct route from start to reward (see Figure 4.1b). In the first maze, the rat enters at A and learns to take route C to the reward. In the second maze, route C is still available, but a new route (*) will lead the rat directly to the reward. If spatial memory were a simple association between stimulus and response, the rats would be expected to choose C in the second maze. But they prefer the more direct route instead. They can construct efficient detours, shortcuts, and novel routes to the reward. These experiments, and others like them, suggest that spatial memory involves the formation of an internal spatial representation—a cognitive map—by which different locations and directions in the environment can be assessed. This characterization of the phenomenon guided neuroscientists to search for a localized region of the central nervous system (CNS) that could serve as a representation of space.

PHENOMENA ARE NOT JUST PROPER FUNCTIONS
Presumably the ability to locate one's self and others in space is an advantage to most creatures, and no doubt the appearance of such abilities conferred on the

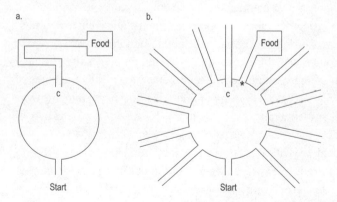

Figure 4.1 Mazes demonstrating the existence of a spatial map. Rats are first trained on maze a. They are then placed in maze b. If learning were a simple pairing of stimulus and response, they would be expected to choose the central route, C. Instead, they take the more direct route, marked *.

organisms that have them significant advantages over organisms that lacked such abilities. Not all phenomena are evolutionarily advantageous. That is, not all phenomena are proper functions of some part of the organism. The phenomenon in question need not do anything beneficial for the organism; indeed it might be harmful. Many biologists study disease mechanisms. Some disease mechanisms involve the malfunction of mechanisms that underlie functions. Parkinson's disease, for example, is thought to be caused by the death of neurons in the basal ganglia, the structures of which are working entities in mechanisms of motor control. Other disease mechanisms cut across the proper functional mechanisms operative in an organism. Although the HIV virus is thought to attack the immune system, for example, in doing so it has secondary effects that cut across a variety of other physiological and cognitive mechanisms. Sometimes the phenomena to be explained are laboratory effects that are of little or no significance to the organism in its natural environment; they occur when the organism is put in conditions that would never or rarely be encountered in the normal operating conditions of the mechanism. A biologist might even be interested in the mechanisms by which his or her techniques produce a given artifact or mechanisms that might be commandeered for some practical application. Not all of the phenomena for which biologists seek mechanisms are effects that have been selected for their contributions to fitness during evolutionary changes.

PHENOMENA, DATA, AND EXPERIMENTAL TECHNIQUES
Phenomena in this sense should not be thought of as data. Data provide evidence for the existence of phenomena. Data in the sense intended here are idiosyncratic to particular experimental arrangements; phenomena, in contrast, are expected to be detectable in a variety of experimental arrangements. Different experimental arrangements reveal different aspects of the phenomenon or reveal the same aspects of the phenomenon in different ways (just as temperature might be measured with alcohol, mercury, and touch). Phenomena are typically the kinds of things that potentially can be detected, manipulated, or produced in many ways across different experimental arrangements or observed with a wide variety of observational methods.

Except when noted, we will speak of the phenomenon generally as a repeatable type of event or product. However, we do not thereby intend to exclude the explanation of particular events by particular collections of entities and activities organized together into particular, unrepeatable mechanisms. Some phenomena of interest to biologists are unique historical events, such as the origin of the 1918 influenza virus or the evolutionary origins of color vision. In such cases, it is less important that the phenomenon be repeatable than that it should be detectable, observable, or otherwise accessible from diverse independent experimental and theoretical perspectives. Otherwise, one doesn't know much about the phenomenon to anchor the search for its mechanism.

One's characterization of a phenomenon at a given time is shaped crucially by the accepted or available experimental protocols for producing, manipulating, and detecting it. Mazes of varying complexity led Tolman to think of spatial memory in terms of the formation of spatial maps. One of the many ways spatial memory is tested is in the Morris water maze. This water maze is a circular pool filled with an opaque liquid. Mice are trained to escape onto a submerged platform hidden somewhere in the pool over repeated trials. They do not like to swim, and so they learn quickly. The liquid is also important because it eliminates smell as a sensory cue that might otherwise explain how the mouse learns the way to the platform. The maze thus isolates the role of visual information in spatial memory. The Morris water maze is a task that allows scientists to investigate the phenomenon of spatial memory, but the Morris water maze task should not be confused with the spatial memory phenomenon itself. Spatial memory is the sort of general phenomenon that might be investigated with a variety of different experimental tasks (such as Tolman's mazes, radial arm mazes, and dead reckoning tasks).

Scientific disputes sometimes turn on the appropriateness of a given experimental arrangement for producing, manipulating, or detecting a given phenomenon. Techniques are sometimes criticized on the ground that they are physiologically irrelevant (that is, that they presuppose conditions relevantly different from those seen in a typical, healthy organism or system of the relevant type) or ecologically irrelevant (that is, they place the organism or system in conditions that they would rarely face outside of a laboratory setting). Some researchers have criticized certain uses of the Morris water maze on these grounds. For example, the Morris water maze has been used to assess learning and memory in mice despite the fact that mice (unlike rats) are not especially aquatic organisms. Mice also have oils in their fur that make them especially buoyant, and so some of them simply float around until the trial period ends. Others turn in circles or swim to the nearest side of the tank and sit motionless until some kind researcher plucks them from the milky water. Debates over the physiological validity or the ecological appropriateness of a task are debates over the character of the phenomenon.

A dramatic example of the mismatch between a single task and the phenomenon under investigation is Bal Krishan Anand (1917–2007) and John R. Brobek's (1914–2009) work in the 1950s on the role of the lateral hypothalamus, a central and primitive brain region, in hunger and feeding behavior. They found that rats with lesions to the lateral hypothalamus stop eating. Later researchers confirmed these results but showed that the rats also stop moving and, in fact, stop responding to most perceptual stimuli. The failure to approach the phenomenon from multiple experimental angles may encourage a most misleading characterization of the phenomenon. A limited task (measures of food intake) is taken to represent the phenomenon (hunger), but the real phenomenon (general arousal) is obscured by the choice of experimental task. The choice of experimental protocol for measuring a phenomenon, in other words, can often lead the researcher to a myopic view of the subject matter. For this reason, any given experimental technique should be seen as providing only a partial view of the phenomenon under investigation, if in fact the technique even taps the phenomenon at all.

WHAT IS INVOLVED IN CHARACTERIZING A PHENOMENON?

Clearly the characterization of a phenomenon is crucial for thinking about the mechanisms that might possibly explain it. Mechanisms and phenomena are reciprocally related to one another in that the character of the phenomenon constrains what will count as an adequate explanation for that phenomenon. Conversely, findings about the mechanism can force one to revise the characterization of the phenomenon (as we discuss below).

One way to tell that a description of a mechanism is sketchy and incomplete is that it explains only some aspects of the phenomenon and not others. If a hypothesized schema fails to explain all aspects of the phenomenon in which the investigator is interested, or if it makes predictions that do not accord tolerably well with aspects of the phenomenon as it is observed in the wild (or, more generally, under the conditions in which one is interested in explaining the phenomenon), then it is not an adequate description of a mechanism for the purposes at hand. A mechanism includes all and only the entities, activities, and organizational features relevant to the phenomenon to be explained. Anything less is, in some sense, incomplete; anything more includes something irrelevant.

In what follows, we discuss different aspects of the characterization of a phenomenon, including precipitating conditions, manifestations, inhibiting conditions, modulating conditions, nonstandard conditions, and by-products. We discuss each of these aspects of the phenomenon with an eye towards how it constrains the space of possible mechanisms for that phenomenon.

PRECIPITATING CONDITIONS AND MANIFESTATIONS

Part of characterizing the spatial memory phenomenon involves noting that the phenomenon is produced or comes about only under a given range of *precipitating conditions*. Some speak of the precipitating conditions as inputs, stimuli, or triggers. We include among the precipitating conditions all of the many sets of conditions sufficient to make the phenomenon come about (or, in stochastic cases, to make the phenomenon as likely as it is). Typically, scientists focus on a few key factors. When one places a rat in a maze and supplies the rat with a reward when it successfully navigates the maze, one has created the conditions for the acquisition of spatial memories. It is an empirical matter, of course, which conditions are required for an organism to begin to engage spatial learning mechanisms and which are required for the spatial learning system to work. In order to fully understand the mechanism of spatial memory, one should at least know the key conditions for the phenomenon to become manifest.

Similarly, one should characterize what it means to manifest the phenomenon. It is too much to demand that every concept in science have a precise operational definition, namely a set of test conditions that exhaustively define the phenomenon. It is nonetheless true that there must be some experimental means of determining whether or not the phenomenon in question is present or absent at a given time. Phenomena are in part specified by the kinds of techniques and experimental strategies for assessing them. In the case of spatial learning, different kinds of spatial learning (such as dead reckoning and navigation by landmarks) are studied using different mazes. Presumably, however, the phenomenon can be characterized in terms of features that can be understood independently of any particular method or technique: phenomena have different durations, time courses, magnitudes, polarities, energy requirements, frequencies, and so on. Learning takes time, as described by learning curves, and fades with time, as described by forgetting curves. Some kinds of learning require one exposure and last a lifetime. Others (such as a cigarette habit) require repeated exposure and can be frustratingly recalcitrant to extinction.

The precipitating conditions and manifestations of phenomena are, as these brief remarks suggest, typically multifaceted and not susceptible to any tidy description. The precipitating conditions are the set-up conditions of the mechanism. Many are listed in the methods sections of research papers investigating the phenomenon.

INHIBITING CONDITIONS

Further constraints to be used in constructing, evaluating, and revising a mechanism schema can be discovered by investigating the inhibiting conditions for a phenomenon—that is, the conditions under which the phenomenon fails or is blocked from occurring. For example, animals with damage to a brain structure known as the hippocampus (on both sides of their brains) cannot perform spatial learning tasks. One can also inhibit learning with genetic modifications to NMDA receptors (protein channels in hippocampal neurons that bind to the neurotransmitters glutamate and glycine). One can also inhibit spatial learning with antagonists to the neurotransmitter glutamate, with inhibitors of protein synthesis, and with high extracellular concentrations of magnesium. The surprisingly elegant explanation for all of these inhibitory conditions is that spatial learning depends on NMDA receptors, which are primarily responsive to glutamate, are blocked by magnesium ions, and which initiate a biochemical cascade that requires protein synthesis in order, ultimately, to restructure the synapses in ways that change the magnitude of its response to the same concentrations of glutamate. Researchers trust this schema in part because it so nicely makes sense of the variety of inhibitory conditions on the mechanism. If one understands the mechanism of spatial memory completely, one should be able to say why spatial memory fails when it fails. A schema that cannot make sense of such features of the phenomena is, *ipso facto*, suspect.

MODULATING CONDITIONS

A complete characterization of the phenomenon also requires knowing the *modulating conditions* for the phenomenon—that is, knowing how variations in conditions change the way the phenomenon is manifest. Spatial learning can be facilitated or inhibited in many ways: by changing the number of learning trials, by changing how those trials are grouped, and by changing the nature of the stimuli used for training. One can also alter a subject's performance on most kinds of memory tasks by changing the retrieval method by which it is assessed, for example, by using free recall, cued recall, or recognition paradigms. A complete understanding of a given type of memory should explain why performance differs under these modulating conditions. If a schema cannot do so, it is lacking in this respect.

NONSTANDARD CONDITIONS

Typically a characterization of a phenomenon includes not just the conditions in which it regularly or naturally or normally occurs but also unusual, or nonstandard, or laboratory conditions. For example, one might be interested in understanding spatial memory only under conditions present in the wild. For the purposes of doing science, however, it is frequently necessary to remove the system under study from its natural or normal context and to study how it behaves under highly controlled and even unnatural conditions. One might be interested in testing, for example, a dose-response relationship between a neurotransmitter and the behavior of a cell and be interested in concentrations of neurotransmitters well below or above levels typically seen in the living organism. The behavior of a mechanism under such unusual conditions is often revealing. Memory systems, to mention another example, are as interesting for what they cannot do (the limits of human memory) as they are for what they can do. Humans rarely encounter nonsense syllables in their day-to-day lives, but, in the wake of Herman Ebbinghaus's (1850–1909) pioneering studies of human memory in the late nineteenth century, nonsense syllables became the standard way to test recall. Staying within this experimental paradigm, one is interested not just in the ability to remember nonsense syllables (e.g., bok and nef) within normal human memory range, but what happens when lists go well beyond that range (e.g., people tend to remember items from the beginnings and ends of such lists better than items in the middle). Going further, one might try to vary the rate at which list items are presented, perhaps exceeding rates people typically experience in their day-to-day lives.

Unless one knows how the phenomenon manifests itself under a variety of *nonstandard conditions*, one does not fully understand the phenomenon. Two how-possibly mechanisms might account equally well for the capacity of a rat to learn how to run a maze under environmentally normal precipitating conditions. However, they may diverge considerably in their ability to account for features of maze-learning that show up in the inhibiting, modulating, and otherwise nonstandard conditions that are so frequently the basis of experimental design. Nonstandard tests are frequently the sieve through which how-possibly is sorted from how-actually.

BY-PRODUCTS

A variety of *by-products* or side effects of the phenomenon can also be crucial for sorting how-possibly from how-actually models, as well as distinguishing incomplete sketches from complete mechanistic models leaving no black or gray boxes to be filled. By-products include a range of possible features that are of no functional significance for the phenomenon (for example, they do not play any role in a higher-level mechanism), and that may even be failures of the system, but are nonetheless crucial for distinguishing mechanism schemas that otherwise account equally well for the phenomenon. For example, the performance of some cognitive tasks competes with the performance of other cognitive tasks (just as rubbing one's head and patting one's tummy can interfere with one another) and fails to compete with different cognitive tasks (just as patting one's tummy and whistling "The Internationale" need not compete with one another). Such interference effects are presumably byproducts of the mechanism's operation and are not part of how it works. However, the fact that it interferes with another task can be very informative in searching for a mechanism. Such interference, for example, might suggest that the two tasks share a common component.

In other cases, the idea of a byproduct is quite literal. One might, for example, learn that a given mechanism gets its energy from ATP by detecting the accumulation of orthophosphate as it works. Or one might sort various hypotheses about the mechanisms gating the sodium channel in neurons by determining which of them can account for the observed gating charge (the movement of a positive charge from inside to outside the membrane) as the channel opens. Again, such byproducts are not working parts of the mechanism, but they provide crucial clues about how the mechanism works.

SUMMARY: CHARACTERIZING THE PHENOMENON
One can conjecture a mechanism that adequately accounts for some narrow range of features of the phenomenon, but that cannot account for the rest. For this reason, descriptions of the multiple features of a phenomenon, of its precipitating, inhibiting, modulating, and nonstandard conditions, and of its byproducts, all constrain the search for mechanistic explanations and help to distinguish how-possibly from how-actually explanations. Similarly, mechanism sketches, with large gaps and question marks, may explain some aspects of the phenomenon but fail to explain others.

RECHARACTERIZING THE PHENOMENON
The search for a mechanism can fail because one has mischaracterized the phenomenon. Three kinds of mischaracterizations are (1) claiming that a phenomenon exists when there is none, (2) lumping together two separable phenomena produced by different mechanisms, and (3) incorrectly splitting one phenomenon into many. Consider each of these ways of mischaracterizing the phenomenon.

FICTIONAL PHENOMENA
One way that the search for mechanistic explanations can fail is by trying to explain a fictional phenomenon. Contemporary spatial memory researchers, for example, have not set out to explain how it is possible, as Plato (429–347 BC) suggested, for creatures to recover lost knowledge of the Platonic Forms from the life before their birth. They are not, to our knowledge, interested in why Oedipal wishes are especially prone to repression in memory, as proposed by Sigmund Freud (1856–1939). They do not seek to understand the formation of L. Ron Hubbard's (1911–1986) engrams or the role of supposed engrams in humans' reactive minds. Clearly it is important for fruitful science that the

phenomenon for which one seeks a mechanism exists. In scientific contexts (and elsewhere), a central test for such existence is whether the phenomenon in question can be detected with multiple causally independent techniques and, furthermore, detected under conditions in which the target mechanism works.

LUMPING AND SPLITTING

Slightly less obvious are the diverse *taxonomic* errors that one might make in characterizing the phenomenon. If the goal is to provide a mechanistic explanation, the phenomena should be chunked in such a way that they correspond to distinct underlying mechanisms. A single phenomenon might correspond to a single underlying mechanism or to the coordinated behavior of many distinct sub-mechanisms. Often, a biologist's understanding of a phenomenon coevolves with her understanding of underlying mechanisms. This coevolution frequently involves recharacterizing the phenomenon in order to bring one's understanding of the phenomenon into alignment with what one is learning about how the underlying mechanisms work.

For example, in a *lumping* error, one might assume that several distinct phenomena are actually one, leading one to seek out a single underlying mechanism when one should in fact be looking for several more or less distinct mechanisms. One crucial conceptual advance in the science of memory has been the effort to identify several distinct types of memory (such as echoic memory, working memory, episodic memory, semantic memory, classical conditioning, and so on) and different types of memory processes (such as encoding, storage, and retrieval). These different types of memory are also in many cases thought to be localized to different regions in the brain and to involve distinct mechanisms. They can be disrupted by different kinds of background and precipitating conditions; they exhibit different learning curves, different forgetting curves, and so on. The fact that organisms have been tinkered together during evolution gives one some reason to expect that different systems of the body (including subcellular systems) will be more or less modular, or (to use Herbert Simon's phrase) "nearly decomposable" in this sense. Although it sometimes takes some idealization to separate out distinct mechanisms, they are often modular enough that it makes sense to split what was previously thought to be a single phenomenon into many.

Conversely, one might commit a *splitting* error, which involves assuming that one phenomenon is actually many. For example, rusting, burning, and breathing were once taken to be different phenomena with different mechanisms rather than different expressions of a common oxidation mechanism. Taxonomic errors are not always confined to a single phenomenon; sometimes they infect entire taxonomies. Franz Joseph Gall (1758–1828), for example, believed that philosophers were wrong to explain the mind in terms of such "mere abstractions" as action, memory, perception, cogitation, and will. Gall's organological system, in contrast, is tailored to identify the set of talents that might vary from individual to individual. His map of the skull contains cranial regions dedicated to the instinct to murder, tenderness for one's offspring, mechanical skill, facility with colors and coloring, and the impulse to propagation. Contemporary cognitive scientists have a different taxonomy. They divide the mind into such phenomena as motion detection, working memory, pitch perception, and computation of expected reward. The point of this comparison is that it is possible that an entire taxonomic system could be ill matched to the mechanistic structures being studied. If so, the taxonomic system is clearly not suited to the search for mechanisms. In short, discovering mechanisms requires not merely picking a phenomenon that is real, but picking one that, so to speak, cuts the world at its mechanistic joints.

In summary, the search for a how-actually mechanism schema can fail because one has tried to explain a fictitious phenomenon, because one has characterized the phenomenon incompletely, or because one has characterized it incorrectly.

CONCLUSION

The search for mechanisms must begin with at least a rough idea of the phenomenon that the mechanism explains. A complete characterization of a phenomenon details its precipitating, inhibiting, and modulating conditions, as well as noting nonstandard conditions in which it can (be made to) occur and any of its by-products. During discovery episodes, a purported phenomenon might be recharacterized or discarded entirely as one learns more about the underlying mechanisms. Lumping and splitting are two common ways of revising one's characterization of the phenomenon in the search for mechanisms.

BIBLIOGRAPHIC DISCUSSION

Bechtel and Richardson (1993) first discussed the importance of "reconstituting the phenomena" and various ways of doing so in light of findings about the underlying mechanism as mechanistic research programs develop. Many of their ideas are summarized in parts of this chapter. The idea that characterizations of phenomena "coevolve" with descriptions of mechanisms traces for us to Patricia Churchland's (1989) discussion of the relationship between psychology and neuroscience, to Schaffner's (1993) and Hooker's (1981a, 1981b, and 1981c) discussions of reduction.

For more on the distinction between data and phenomena, see the classic article in philosophy of science on the subject, Bogen and Woodward (1988) and replies to criticisms in Bogen (2011) and Woodward (2011).

On the relation between the sense of "function" defended here and the "proper" sense of function, see Robert Cummins (1975). The sense of role-function that we defend is developed in Craver (2001) and Craver (forthcoming). Arguments in favor of an etiological view of proper functions (i.e., a function that arose via natural selection) can be found in Larry Wright (1973; 1976) and Karen Neander (1991). For reviews of issues about functions in biology, see Colin Allen et al. (1998) and Arno Wouters (2005).

For work on modularity and near decomposability, see Simon (1996). For a more recent discussion from the perspective of mechanisms, see Steel (2008).

Plato discusses his doctrine of recollection in his dialogues the *Meno, Phaedo*, and *Phaedrus*. Edward Tolman's classic studies of maze learning can be found in Tolman (1948) and Tolman and Honzick (1930). For more on the use of different mazes to isolate different aspects of spatial learning, see Chapuis et al. (1987) and Olton and Samuelson (1976). For more on memory systems and their separation, see Schacter and Tulving (1994) and Roediger (2008). On the hypothalamus as a feeding center, see Anand and Brobeck (1951).

Herman Ebbinghaus's collection of research on the structure of human memory, including the primacy and recency effects discussed in this chapter, can be surveyed in Ebbinghaus (1885).

The classic statement of operationalism, that "we mean by any concept nothing more than a set of operations; the concept is synonymous with the corresponding set of operations" is due to Bridgman (1927, p. 5). See also Hasok Chang's (2009) entry on "Operationalism" in the *Stanford Encyclopedia of Philosophy* and his discussion in *Inventing Temperature: Measurement and Scientific Progress* (2007).

The discussion of lumping and splitting borrows significantly from the work of geneticist Victor McKusick (1921–2008); see his (1969) article on lumpers and splitters. For further discussion of lumping and splitting, see Craver (2004; 2009). For an application of this discussion to the categorization of psychiatric disorders, see Kendler, Zachar, and Craver (2011).

5 STRATEGIES FOR MECHANISM SCHEMA CONSTRUCTION

DISCOVERY: FROM AHA TO STRATEGY

In folklore, scientists are often represented as fortunate shamans. Discoveries strike them in mysterious flashes delivered, it would seem, from a realm beyond reason. Archimedes (287–212 BC), legend has it, screamed "Eureka" at the moment the law of buoyancy popped into his head. Isaac Newton (1642–1727), the story goes, was struck (figuratively) by the idea of gravity when his head was struck (literally) with an apple. August Kekulé (1829–1896) claimed to have discovered the ringed chemical structure of benzene as he drifted to sleep while watching the sparks from a fire whirl above his head; in the more common version of the story, Kekulé's discovery struck him as he dreamed about a snake swallowing its tail. Mysterious, indeed.

Let us grant that this aha perspective captures something familiar even in the mundane discoveries of everyday life. Everyone has a story about the problem that foiled her time and again until suddenly, as if by magic, the solution appeared full-blown before her mind's eye. Scientists have those experiences too. Perhaps precisely because such stories are so compelling and familiar, however, they tend to obscure the frequency with which scientists solve problems by reasoning about them, using strategies to construct hypotheses, to test those hypotheses, and to revise them as necessary. Scientists use construction strategies to imagine how a mechanism might possibly work or to populate the space of possible mechanisms. They use evaluation strategies to identify constraints that prune the space of possible mechanisms. They use revision strategies to diagnose and localize the source of error in a hypothesized mechanism and to direct the search for an amended schema to fruitful regions of the space of possible mechanisms.

Our goal in this chapter is to make some of those strategies explicit in the hope that scientists struggling in a discovery episode might find new ways of thinking about their problems or that students entering the biological sciences will see directly the importance of internalizing the wealth of background knowledge from their own and other sciences. The history of science is a source of compiled hindsight about what has and has not worked in the search for mechanisms. We examine historical cases for strategies and perspectives on discovering mechanisms that a biologist might find useful now.

THE PRODUCT SHAPES THE PROCESS OF DISCOVERY

Our topic is not discovery in general, but the discovery *of mechanisms* in particular. The nature of the product—a mechanism schema—shapes the process by which it is discovered. The goal is to find a set of entities and activities, to describe how they are organized together, and to show that when they are organized together just so, they produce the phenomenon one is trying to explain. One is not merely seeking an economical equation that describes the phenomenon. One is not seeking merely a set of correlated variables. One is not seeking the function that a given phenomenon serves in some context. Rather one is attempting to construct a mechanism schema that describes how components are organized together to do something.

This goal, the search for mechanisms, shapes the process of discovery. The idea that one is searching for mechanisms leads one to seek component parts, their activities and interactions with one another, and features of their organization. The idea that one is searching for a mechanism allows one to work on the discovery project in a piecemeal fashion; one can work on one part of the mechanism at a time while leaving other parts as black or gray boxes. And because one is attempting to reveal the productive continuity of a mechanism from beginning to end, what one learns about one stage of the mechanism places constraints on what likely has come before or what likely comes after a given stage. Whenever we are looking for a mechanism, we are driven to ask: What are the set up and finish conditions? Is there a specific, triggering start condition? What is spatially next to what? What is the temporal order of the stages? What interacts with what? What are the specific entities and activities in the mechanism? How does each stage of the mechanism give rise to the next? How was each stage driven by the previous one?

Similarly abstract guidance comes from the decision about whether one is seeking a mechanism that produces, maintains, or underlies a phenomenon. The three diagrams in Figure 5.1 show very abstract schemas for these three cases. These abstract diagrams can be filled with appropriate entities, activities, and organizational features (e.g., feedback loops, spatial separation, etc.) to yield a mechanism schema with sufficient details for the problem at hand. In the case of a productive mechanism, one typically starts with some understanding of the end product and seeks the components that are assembled and the processes by which they are assembled and the activities that transform them on the way to the final stage. In the case of an underlying

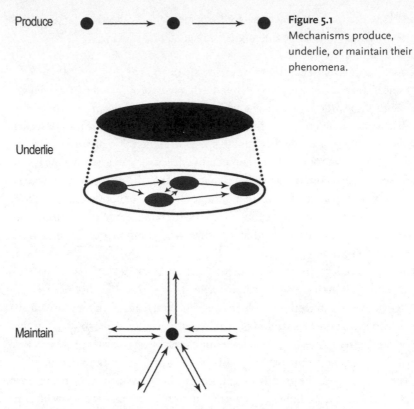

Produce

Figure 5.1
Mechanisms produce, underlie, or maintain their phenomena.

Underlie

Maintain

mechanism, one typically breaks a system as a whole into component parts that one takes to be working components in a mechanism, and one shows how they are organized together, spatially, temporally, and actively such that they give rise to the phenomenon as a whole. One need not know from the start how these facts about mechanisms and their organization fit together to begin learning what one can about the likely components, their properties, and their relations to one another. Finally, in the case of a maintenance mechanism, one typically needs to characterize some process or property (the homeostatic set point, shown in the center of the diagram) that is maintained at a given speed or level, one needs to recognize the forces that tend to move the system away from its homeostatic set point, and one needs to characterize the process by which those divergences are detected and/or corrected. The very fact that one is searching for a mechanism that produces, underlies, or maintains the phenomenon thus leads one to ask and emphasize different kinds of questions. Progress in mechanistic sciences consists in the answers to such questions.

A mechanism discovery episode typically begins with at least two sources of empirical knowledge to guide the discovery process. The scientist comes to a problem with a store of knowledge about the phenomenon under investigation and about the target system in question. First, one begins knowing something, even something sketchy, about the phenomenon for which one is hoping to discover a mechanism (see Chapter 4). Second, one typically begins with prior knowledge about the kinds of entities and activities that might be involved in such phenomena in a given type of organism or system. To describe a phenomenon is to characterize it in the language of a given field and to implicitly call up the host of explanatory concepts, the store of accepted entities, activities, and organizational structures that people in the field are licensed to use in constructing a mechanism schema.

If the phenomenon is of a familiar type, for example, the field might already have developed a store of mechanism-types through which one might search. Table 5.1 illustrates resources that one might consult in attempting to discover a mechanism for a given phenomenon. For example if one begins by trying to understand how two cells or two populations of cells communicate with one another, then there are three known kinds of communication one might consider. If the cells are in direct contact with one another, then one might consider the possibility of a juxtacrine form of communication, such as an electrical gap junction (which directly connects the cytoplasm of one cell with the cytoplasm of another) or an interaction among cell-surface proteins (as occurs in certain mechanisms of cell differentiation, such as the Notch signaling mechanism). If the cells are not in direct contact but relatively close to one another (such as between a pre- and postsynaptic cell in the visual cortex), then one might consider paracrine mechanisms, such as the release of neurotransmitters. If on the other hand one is dealing with long-distance communication (such as between the pituitary gland in the brain and the kidneys) or the control of many systems by a central and distant regulatory mechanism (such as the pituitary gland communicating at once with the veins, the kidneys, and the heart) then one might consider endocrine forms of communication. Once a choice among these has been made, further choices can be considered. If one is looking at a form of paracrine communication among neurons, for example, then one might consider a list of known neurotransmitters, perhaps narrowing the search to those that are known to be found in a given brain region of interest. In short, one often begins the search for mechanisms with a ready-made layout of the space of possible mechanisms, and the goal is to use empirical findings to eliminate regions of that space. The nature of the phenomenon and the store of accepted

Some Types of Mechanisms	
Adaptation Mechanisms	Locomotion Mechanisms
Selection	Burrowing
Instruction	Flying
Preorganization	Swimming
Articulation Mechanisms	Terrestrial
Ball and Socket	Jumping
Hinge	Slithering
Pivot	Walking
Saddle	Reproduction Mechanisms
Cellular Transport Mechanisms	Sexual
Osmotic, e.g., diffusion	Anisogamous
Passive, e.g., glucose permease	Isogamous
Active, e.g., Na+/K+ pump	Asexual
Cell Signaling Mechanisms	Binary Fission
Juxtacrine	Budding
Electrical	Conjugation
Chemical	Fragmentation
Paracrine (short distances)	Parthenogenesis
Endocrine (long distances)	Spore Formation
DNA Repair Mechanisms	Synthesis Mechanisms
Mismatch Repair	Joining
Base or Nucleotide Excision	Dividing
Repair	Copying
Double Strand Break Repair	Templating
Gene Regulation Mechanisms	Assembly
Gene-1 regulates Gene-2	
negative or positive	
Autoregulation	
negative or positive	
Oscillating Cycles	

Table 5.1

mechanism types thus work together to frame the discovery problem: to reveal the layout of the space of possible mechanisms and perhaps to tell one how to decide among them.

An engineer faced with the task of inventing a new device to perform a task begins by brainstorming the available options. The nature of the task, the distances involved, the time urgency, and the price—all place limits on the space of plausible solutions. For the biologist in search of mechanisms, different limits are in place: the space of possible mechanisms is limited by the space of available parts and activities in the target biological system at a given time. *The space of live possibilities* is thus quite a bit more restricted than the space of possibilities more generally. An engineer might contrive any number of distinct artifacts, cobbled from altogether different components, to do one and the same job; they are not constrained by the parts that evolution and genetics make available in a given kind of organism at a given time. Among biological systems, however, the options are constrained by the parts of our ancestors, by the developmental mechanisms by which genes and the environment shape an organism, by constraints on overall fitness. The human eye cannot weigh twenty pounds. An artificial vision device might be arbitrarily heavy. The point is that biologists typically begin their search for mechanisms in an intellectual cornucopia of general knowledge that restricts the space of possible mechanisms and focuses attention on a few or a handful of key live possibilities. More specifically, a discovery episode plays out against many kinds of background knowledge about the available types of mechanisms for phenomena of that sort, about their applicability to the target system, and about the means for testing among them. Students learn these kinds of facts about their domain of study in introductory courses, and experts refine this sort of knowledge over the course of their careers. Although occasionally biology calls upon a Darwin or a Harvey to introduce an altogether novel kind of mechanism, most episodes of mechanism discovery play out within the space of known mechanism types within a scientific field. Characterizing those types is thus of particular pedagogical import; during a discovery episode they might provide useful guidance for thinking about how a new mechanism might work.

LOCALIZATION

Localization is a second, often crucial, clue in the construction stage of the search for mechanisms. It is often far from obvious at the beginning of a discovery episode where in the system under study the phenomenon takes place. In some cases it is unclear whether the phenomenon even takes place in the system under study at all, or whether, in contrast, it takes place in the interactions between the system and its environment (or perhaps entirely in the environment).

The phenomenon of inheritance came to be localized to hereditary materials inside germ cells. Inheritance is the tendency of offspring to resemble their parents. In sexually breeding species, inheritance is biparental; in other words, the child has a mixture of traits from both parents. By the mid-nineteenth century, when researchers discovered the mammalian egg cell, the idea that the sperm and the egg transmitted the hereditary material was quite plausible. It remained controversial, however, whether influences from a previous mating could linger in the mother's womb.

Gregor Mendel's (1822–1884) work in the 1860s on patterns of inheritance, rediscovered around 1900, provided evidence for unobservable differences among germ cells that might explain the preservation of traits. Further evidence for localization of hypothesized hereditary particles within the germ cells occurred after the discovery in the late nineteenth century of chromosomes (threadlike bodies in the nuclei of cells) that were hypothesized to be the hereditary material. Evidence for this localization was strengthened in the early twentieth century when abnormalities in chromosomes were found to correspond to abnormalities in patterns of inheritance of traits. Biochemists noted that chromosomes are composed of two types of chemicals: proteins and deoxyribonucleic acid (DNA). Chemical analysis of DNA in the 1930s, given the sensitivity of the chemical techniques at the time, seemed to indicate that it had a simple repeating structure, whereas proteins had much more complexity. Hence in the 1930s and 1940s, the most plausible location for the hypothesized hereditary particles, the genes, was in the proteins of chromosomes. This view was dispelled with more accurate chemical analysis of the DNA molecule and the discovery of its double helical structure in 1953. The idea that the gene is a linear sequence of bases along a DNA chain localized part of the hereditary process and, in so doing, offered an entry point into the mechanism from which one could conjecture how genes could possibly produce hereditary traits.

In the search for hereditary mechanisms, evidential considerations ultimately localized the search to a particular molecule in particular (germ) cells of adult organisms. In other cases, one is led upwards to recognize a larger system of which the initial target system is a part. Jellyfish osmoregulate (that is, maintain the concentration of their body fluids within narrowly constrained values) by passively matching the concentration of their extracellular environment and by living in very stable osmotic environments. The mechanism of osmoregulation is not inside the jellyfish; indeed, the jellyfish doesn't do much at all to maintain its osmotic balance. Likewise, researchers studying gene regulation and epigenetics have discovered that much about the development of traits and the expression of genes depends upon behaviors of the organism in their environmental and social context. For example, the capacity of E. coli bacteria to metabolize lactose (the sugar in milk) depends upon the presence of lactose in its environment. In the absence of lactose, the enzymes for metabolizing it are not produced. The mechanism of lactose metabolism, it would seem, reaches beyond the bacterium itself to include an inducer from the bacterium's environment.

Locating (and relocating) the mechanism is often a crucial turning point in discovering a mechanism. By choosing a locus for the search for mechanisms, one places the mechanism in a new context, one avails oneself of vocabularies for describing entities and activities in that locus, one circumscribes the experimental and observational strategies that will be useful in addressing the problem, and one often makes available a set of mechanism schemas that might possibly be brought to bear in a given case.

EMPLOY A SCHEMA TYPE

In the construction phase of the search for mechanisms, it is often useful to consult a store of potential analogies and a store of known types of mechanisms.

Consider analogy first. The nature of the phenomenon may aid in retrieving a specific analogue that furnishes the overall structure for the target schema. The strategy of reasoning by analogy involves the following reasoning processes: retrieval, mapping, adjustment, and evaluation. The nature of the phenomenon provides clues to retrieve an appropriate analogue. After finding an analogue, the scientist then maps aspects of the analogue to the target area and adjusts their fit. The new schema can then be evaluated in light of empirical evidence.

Sometimes useful analogies are, at first, rather distant from the target phenomenon. Early geneticists analogized genes to beads on a string in order to construct a schema to explain inheritance. Some traits were found to be correlated with one another through successive generations; such traits are said to be linked in inheritance (rather than independently assorted, as Mendel assumed traits to be). The geneticist T. H. Morgan (1866–1945) and his colleagues explained the phenomenon by proposing that traits linked in inheritance are due to genes linked along chromosomes. They conceived of chromosomes as linear threadlike bodies in the nuclei of cells. The linear nature of the chromosomes suggested an analogy: chromosomes are like strings, and genes are like beads on the string. One would therefore expect that traits with genes together on a single chromosome would be inherited together.

However, the phenomenon is more puzzling than this simple analogy suggests. Linkage is not complete; occasionally traits that are commonly linked together are nonetheless inherited independently of one another. Morgan had read the work of a microscopist who observed chromosomes lying across one another on slides. But the techniques of the early twentieth century did not allow observation of the dynamics of their behavior in cells. Nonetheless, the analogy of beads on a string suggested the mechanism of crossing over. Parts of the strings cross over, break, and rejoin. The analogy allowed mapping from possible mechanisms for recombining and joining beads on a string to a schema for the mechanisms that produce and break linkages between traits.

Charles Darwin famously used analogies to reason about evolution. As one can read in his notebooks, Darwin early on considered and rejected the analogy between selective breeding of domestic animals and the formation of new species. He could find nothing in nature to play the role of the choosey breeder. Then, in 1838, he read a copy of Thomas Malthus' (1766–1834) *Essay on Population*, for amusement, he tells us. In that work, Malthus predicts that the human population will eventually outstrip its food supply, leading to a struggle for life. Darwin quickly generalized this struggle to all living things. Such a struggle for existence, he reasoned, could play the role of the selector: those organisms with variations advantageous in the struggle tend to survive. Just as the breeder scrutinizes the variations in, say, a brood of puppies to select the ones with the most desirable traits, nature scrutinizes (Darwin's own term) natural variations and selects those that are better able to survive the struggle for existence. This analogy occupies the prominent rhetorical space of the first chapter in Darwin's *Origin of Species* (1859). Analogies often suggest new terminology for describing the target mechanism; however, the language may need to be adjusted to remove failures of fit between the analogy and its target. Some argued that Darwin's language was too anthropomorphic and seemed to personify nature as a selector. Herbert Spencer (1820–1903), for example, suggested that the theory of natural selection could be better expressed as "survival of the fittest." Spencer emphasized that this phrase helps to understand natural selection in mechanical terms. Darwin adopted Spencer's phrase and included it as a chapter subtitle in the fifth edition of the *Origin of Species* in 1869. Darwin thus made adjustments to remove a misleadingly anthropocentric component introduced into the schema of natural selection through his use of the breeder analogy.

The use of analogies during hypothesis construction is well documented. Whether such analogies count as evidence in favor of how-possibly schemas has been a matter of debate. Surely the analogy offers some measure of plausibility: the type of mechanism is already known to exist and to produce analogous effects. However, additional evaluative strategies must clearly be used to assess whether the newly constructed schema applies in the case at hand. (We discuss these in following chapters.)

Rather than (a) reasoning directly from the analogue to the target, another method for constructing a target schema is (b) categorizing the phenomenon as of a certain type and using that categorization to retrieve one or more abstract schemas appropriate to phenomena of that type. Consider some examples.

Claude Bernard (1813–1878) emphasized repeatedly that values of the central parameters required for living and proper functioning must be maintained within narrowly circumscribed ranges against the challenges posed by the external world and changes to the way things work inside the organism. Such homeostatic regulation makes organisms robust across changes in environments and fit to deal with injury and malfunction. When one encounters such tight regulation, one tends to posit the existence of some kind of negative feedback mechanism. They have a common abstract form. A mechanism (called "the plant" in control theory) is operating to produce an output. The output is then used to generate a signal that modifies the input to the mechanism or, by some other means, the behavior of the mechanism itself. For example, dehydration causes animals to drink water and to void salt in urine. These two activities add fluid to and reduce solute levels in the blood and thereby eliminate the dehydration. Likewise, human body temperature exhibits a kind of homeostasis like that seen in household thermostats. A drop in body temperature, for example, signals the body to shiver, which increases body temperature; overheating signals the body to sweat, which decreases the body temperature through evaporative cooling. When one is dealing with mechanisms of control and regulation, feedback of some sort is often involved.

Darwin's interest in evolution was driven in large part by the desire to understand how organisms came to be so well adapted to their environments. How, one might wonder, did the woodpecker's tongue come be so long that it can pluck insects from the trunks of trees through its long beak? Darwin's theory of natural selection supplies a schema; details added to this schema show how it applies to the woodpecker. The schema for natural selection has several stages. The mechanism begins with a population of variant organisms. The population faces a set of environmental challenges that not all of them can survive. Different members of the population deal with that challenge in different ways. As a result, some organisms derive a differential benefit from their particular variation and so are more likely to replicate or survive than are other variants. Woodpeckers with longer tongues have an adaptive advantage.

A more abstract selection schema has abstract stages: variants arise in a population; the population faces challenges not all of them can survive; those with traits advantageous in the struggle for existence tend to survive and tend to reproduce more offspring with the adaptive variations. An alternative way of specifying this schema allows it to apply to different units of selection, such as genes or groups. Genes vary and produce phenotypic traits; those traits better suited to withstand an environmental challenge result in more of the genes associated with them replicating and thus increasing in frequency in the following generation. Alternatively, groups with, for example, effective communal childcare practices or altruism to other group members, might be more likely to populate a given geographic region in virtue of benefits for the group (perhaps even at the expense of individual organisms). Nothing in the selection schema prevents it from applying to different units: most obviously organisms and groups, but perhaps also to cells, molecules, and behaviors. The point here, however, is to see how employing an abstract selection schema plays a role in the construction of a target schema. When the phenomenon to be explained is an adaptive fit between something and its environment (adaptation), then one might consider a selection-type mechanism.

Fields besides evolutionary biology employ selection schemas. For example, in immunology the clonal selection theory borrows this schema type to explain how organisms defend against invading antigens. The theory posits that the body generates a large population of variant lymphocytes. When an antigen attacks the body, the variants adapted to fighting it increase in number. The question is how this happens. The answer is that the lymphocyte cells have variant reactive sites on their surface that determine whether they react with invading antigens; those reacting with the antigen are activated; more cells with such reactive sites are cloned. These then produce antibodies to deactivate the invading antigen. The selection type schema applies here as well.

Where do schema types come from? One reliable source of such schemas and tools for sorting them from one another is the history of science. One can construct and group such schemas by dropping the details of a specific case. This is how one might understand the relationship between natural selection to clonal selection. Dropping details of genic, organismic, and group selection yields a generic natural selection schema specified in each of these cases; dropping further details yields an even more abstract selection schema that may be instantiated for other selection mechanisms, such as clonal selection in immunology.

In some cases, scientists make use of their knowledge of artifacts, such as computers or clocks, to generate how-possibly schema types. The idea that the brain is a computer, and so traffics in information that can be coded, decoded, and manipulated, remains a central guiding assumption of the field of cognitive neuroscience. The engineering work of building a transatlantic cable for sending telegraph signals (by Lord Kelvin in 1855) provided a model for understanding how neurons propagate electrical signals along the length of their dendrites and axons. So in the history of science, mechanisms in nearby fields and artifactual marvels often provide compelling clues about the kind of mechanism that might produce, underlie, or maintain a phenomenon of interest.

MODULAR SUBASSEMBLY

Employing an entire abstract analogous schema to obtain a how-possibly schema is a conservative strategy: one expects to find roughly the same overall framework in the target case. Such a strategy is limited by already available types of analogues and abstract schemas for the type of phenomenon to be explained. It is more adventurous to put together old parts, modules, in new ways.

This process, *modular subassembly*, involves reasoning about how mechanism components might be combined in surprising ways. One hypothesizes that a mechanism consists of (either known or unknown) modules or types of modules. One cobbles together different modules to construct a hypothesized how-possibly mechanism, guided by the goal of finding modules to fill all the gaps in a productively continuous mechanism. In doing so, scientists draw upon their knowledge of the store of types of entities and activities and modules (i.e., interconnected entities and activities that have a particular function).

Evolution itself often works by copy and edit: copies of genes can be found in mutated form, playing similar or different roles in the same or related organisms. Finding such recurrent motifs has been a powerful tool in discovery in biology. There are various types of receptors, types of neurotransmitters, types of enzymes, and types of gene regulatory components (e.g., inducers, repressors, and other types of transcription factors). When a functional requirement can be specified in the mechanism sketch, then appropriate types of modules can be sought to satisfy it. If, for example, an external inducer coordinately controls a group of bacterial genes, then a repressor module can be added to the schema for gene control, involving, for example, a repressor gene and a repressor protein.

Developmental biologists discuss modularity as a principle in evolution: components are individually modified or conserved in different lineages. One of the remarkable cases of a functional module consists of the *Pax6* gene and a group of related genes. This module plays a role in the formation of eyes in invertebrates such as fruit flies and in vertebrates such as frogs and mice. This module recurs even though the eyes themselves have very different structures (compound eyes in flies and eyes with a single lens in vertebrates). Experimentalists can manipulate such modules to investigate their functional roles. An amazing abnormality results from placing a mouse *Pax6* gene in the leg of a fruit fly. Structures characteristic of the insect eye develop on the leg. Hence, if one is seeking the mechanism of eye development in a new species, knowledge of such types of conserved modules will aid in constructing a plausible target mechanism.

Finding a new type of module opens a new region of the space of possible mechanisms. The 1970 discovery of the enzyme, reverse transcriptase, which copies RNA back into DNA, is such a module.

RNA—via reverse transcriptase→DNA

This reverse transcriptase module plays an important role in retroviral infection. Retroviruses, such as the HIV-AIDS virus, carry this enzyme into the infected cell.

Retroviruses use reverse transcriptase to copy their own viral RNA into DNA. The DNA copy of the viral genome is then integrated into the host genome and is copied as new cells are produced. Integration into the host DNA makes such retroviruses very difficult to attack with drugs.

After this reverse transcriptase module was found to function in the mechanism by which retroviruses copy their RNA back into the host DNA, it then became a module for hypotheses about nonviral mechanisms. The discovery of reverse transcriptase opened up a space of possible mechanisms in which base sequences could be copied and inserted into the DNA, a hitherto unknown activity.

The introduction of this new module, for example, raised the exciting possibility that genetic variation might, in some instances, be produced directly under the influence of environmental instruction. (Note that "instruction" is one of the possible sub-mechanisms for producing adaptation in Table 5.1.) The finding thus gave rise to hypotheses about new, instructive NeoLamarckian mechanisms to explain adaptation. The adaptive trait might be produced directly by the environmental challenge. The adaptive trait provides a benefit in that challenging environment, and, in some instructive theories, that trait is then passed on in inheritance. Such instructive mechanisms, if they exist, would thus produce inheritance of adaptive, acquired characters. Instructive mechanisms require elaborate modules for receiving and appropriately responding to an instruction, as well as for appropriately changing the hereditary material (DNA or RNA in some viruses) that is passed to the next generation. Given the elaborate machinery needed in the modules to detect, transmit, and respond to an environmental signal, it is not surprising that few possible, productively continuous, instructive mechanisms have ever been proposed in the history of biology. (It is worth noting, however, that human learners do respond well to instructions, so mental mechanisms are an exception. Of course, those learned responses are not biologically inherited.)

Jean-Baptiste Lamarck (1744–1829) is usually credited with proposing the first instructive theory of evolutionary change in his 1809 Zoological Philosophy. However, he did not propose a productively continuous mechanism. In his famous example, the giraffe's neck got longer and longer as it reached higher and higher for leaves, stretching its neck, and (somehow) passing stretched necks on to offspring. Working prior to the development of the cell theory, Lamarck proposed a fluid physiology in which fluids would flow into the neck to elongate it. But he had no theory of heredity to explain how this elongation was passed on to offspring. This was a large gap in the productive continuity of his schema. With time, biologists increasingly came to think of such instructive changes as impossible. In the late nineteenth century, biologists not only embraced the cell theory, they also drew a sharp distinction between germ cells, the hereditary link from one generation to the next, and somatic cells, which are not involved in reproduction. Transfer of information from the environment or from somatic cells into germ cell DNA (as discovered in the twentieth century) was deemed to be impossible.

So it is little wonder that the discovery of reverse transcriptase and its activity of copying the sequence of RNA back into a DNA sequence was an exciting new kind of module that could possibly fill a gap in an instructive mechanism sketch. In a controversial 1998 book, *Lamarck's Signature*, Edward J. Steele (a contemporary Australian molecular immunologist) and his colleagues proposed a hypothetical immunological mechanism using reverse transcriptase as an instructive module. They claimed to have found evidence for the phenomenon of inheritance of an adaptive acquired character in mice. When the mother acquired immunity, that immunity appeared to be passed on to her offspring. They proposed that some of the messenger RNAs that produced the antibody proteins in the immune response were captured by supposed endogenous retroviruses found in the mother mouse. The retroviruses then migrated to the mother's egg cells and reverse transcribed the antibody RNA into the mother's germ line DNA. This DNA was then passed on to the baby mouse where the antibodies were produced, giving it the same immunity that the mother had acquired.

Evidence for the phenomenon of inheritance via the mother's DNA of an acquired immunity (as opposed to through the baby's ingestion of the mother's milk) was not confirmed by further testing. However, the point here is that the discovery of a new type of module, reverse transcriptase in viruses, expands the space of possible mechanisms for explaining adaptation, regardless of whether the hypothesis ultimately survives empirical scrutiny. Finding recurrent motifs in biological mechanisms provides a growing store of modules of use in constructing how-possibly schemas.

FORWARD / BACKWARD CHAINING

The final strategy for constructing mechanism schemas we consider involves first learning something about the mechanism or one of its components and then using that knowledge to make inferences about what came before it or what is likely to come after it. In forward chaining one uses the early stages of a mechanism to reason about the types of entities and activities that are likely to be found in later stages. In backward chaining one reasons from the entities and activities in later stages in a mechanism to find entities and activities appearing earlier. With cyclic mechanisms some separable stage can serve as a relative starting point for reasoning about earlier or later stages. Thus, the strategy of forward/backward chaining is likely available to scientists when they know anything, or can conjecture anything, about the specific entities and activities at any stage in the hypothesized mechanism.

Consider some of the ways one might use what is known about an entity or activity in a mechanism to reason about earlier or later entities and activities in the mechanism. Entities engage in activities by virtue of the fact that they have the right properties. We call the properties that make such activities possible the *activity-enabling properties* of an entity. One can often use general knowledge about kinds of properties and their association with particular kinds of activities in biological systems to reason forward about the activities in which the entities can or do engage. Such activity-enabling properties include three-dimensional structure and size, as well as location and orientation. Structures can promote or inhibit the push/pull of geometrico-mechanical activities. Three-dimensional shapes can be open or closed, narrow or wide, exposing or concealing. This taxonomy of shapes is closely tied to the activities in which entities with such shapes can engage. An open entity permits movement through it more or less as its opening is narrow or wide. Entities may also have different kinds of charges, and molecules have valences, both of which affect the kinds of bonding activities in which they engage. So, finding the activity-enabling properties of entities gives clues to the kinds of activities in which they might engage in the next stage of the mechanism.

Conversely, in backward chaining, *activity signatures* are a source of clues as to what came before. When an activity operates, it produces an effect that changes the entities involved. Learning the properties that may have been changed, the activity signatures, allows one to conjecture what happened in a previous stage. For example, if one detects a hydrogen bond in a later stage of a mechanism, then one can reason that a prior stage included polar molecules, with complementary weak charges that have been neutralized in the formation of the bond. In another example, consider an allosteric molecule. An allosteric molecule changes shape (via stresses and strains of geometrico-mechanical activities) when it bonds to an effector molecule. The new shape exposes a new active site, so that the allosteric molecule is now able to eject molecules currently bound or to bond to a third molecule. If one detects an allosteric molecule bound to an effector (an activity signature), that indicates that effector bonding occurred in a prior stage.

Forward chaining is illustrated by Watson and Crick's famous suggestion about DNA replication. Watson and Crick close their classic 1953 paper on the double helix structure of DNA as follows: "It has not escaped our attention that the specific [base] pairing we have postulated immediately suggests a possible copying mechanism for the genetic material" (Watson and Crick 1953, p. 737). DNA has polar bases that hold the two helices together with their complementary hydrogen bonds. These entities could obviously play a role in the first stage of a copying mechanism. The double helix could open and allow complementary bases to line up along it in a specific order. The polar charges and their spatial arrangements are activity-enabling properties. Based on these features of the mechanism, Watson and Crick were led to conjecture how the next stage of the mechanism might unfold. Continuing to forward chain, one then could see how two identical helices would result. Of course, this conjecture is now the central pillar of our understanding of heredity.

Backward chaining and forward chaining both contributed to the discovery of the mechanism of protein synthesis in the 1950s and 1960s. Biochemists knew that the endpoint of the protein synthesis mechanism was a string of amino acids held together by strong covalent bonds. They thus reasoned back to an antecedent stage in which the amino acids were free and unbound. Because energy is required to form such strong bonds, reasoning backward suggests the existence of a high energy intermediate in the immediately preceding step. Biochemists isolated such a high energy intermediate. Thus, the strong covalent bonds were activity signatures, indicating components needed in the preceding steps to form them. Surprisingly, the activated amino acid was associated with RNA. A typical biochemical reaction schema to synthesize a molecule has the separate chemical components on one side and the newly synthesized molecule on the other. Accordingly, biochemists expected amino acids and an energy source to be required for synthesizing a protein (a chain of amino acids). However, such a schema has no role for RNA because RNA is not a component of proteins. Reasoning backward from protein to free amino acids did not suggest an RNA intermediate in the chemical reaction.

Meanwhile, molecular biologists were reasoning forward from the DNA double helix to the next stage in the protein synthesis mechanism. Biochemists and cell biologists had discovered that RNA was involved in the mechanism. Molecular biologists suggested that RNA acted as a template that would guide the assembly of the protein. The order of the bases in the coding strand of DNA would be transcribed into similarly ordered bases in RNA. The RNA then serves as the template that orders the amino acids in the protein. Biologists used these tandem strategies to fill gaps in the productive continuity of the proposed mechanism. The molecular biologists reasoned forward from the DNA, while the biochemists reasoned backward from the finished protein. Their work met in the middle of the mechanism, with the discovery of the various types of RNAs and their roles (as we discuss in more detail in later chapters). One biochemist suggested that the scientists were like workers building a tunnel: they started at opposite sides of a mountain, and eventually they met in the middle.

In sum, forward and backward chaining are reciprocal strategies for reasoning about one part of a mechanism on the basis of what is known or conjectured about other parts in the mechanism. They may be used independently of or in conjunction with the other two strategies discussed above. The strategy of employing a schema is a top down strategy that provides a how-possibly overall organizational structure for the target mechanism. Modular subassembly involves putting together working subcomponents of a schema. Finally, at a finer grain, one can construct a hypothesized mechanism by reasoning forward or backward about the entities or activities themselves.

The use of each of the four types of strategies depends on what conceptual resources are available and what is known about the target mechanism already. As noted above, to employ a type of abstract schema, one needs to note that the phenomenon is a certain type and one must know an analogue or abstract schema for that type of problem. These may or may not be available. To construct a schema via modular subassembly one needs at least to sketch the functional requirements for some stages of the mechanism so that appropriate modules can be sought within the store. To localize a function, one must have some knowledge of the relevant functions and of what might reasonably count as a location. Similarly, to employ forward/backward chaining, one needs to know or conjecture the beginning point for the mechanism, the endpoint, or some cut point in the middle.

CONCLUSION

In this chapter, we discuss how the product shapes the process of discovery, how the search for mechanisms is guided by the very fact that it is a search for mechanisms rather than something else (e.g., an entity, a technique, or an abstract mathematical law). The scientist constructing a mechanism schema is frequently less like a fortunate shaman conjuring a divine message than like an engineer drawing a blueprint for an artifact. The product is ultimately produced through sound training and hard study, which supply an abundant store of components, modules, and schemas on which to draw, and hard intellectual work of imagining how these might be adjusted and combined in the case at hand. Such insights often are born through the diligent application of strategies for constructing mechanism blueprints, such as: localization (to find where the mechanism operates), employing an analogous or abstract schema (to provide the overall organization of the mechanism), modular subassembly (to provide functionally characterized groupings of entities and activities), and forward/backward chaining (to provide what comes next or what comes before any given stage).

BIBLIOGRAPHIC DISCUSSION
On the truth of the Newton apple story, see Gefter (2010). The Kekulé story about the discovery of benzene is discussed in Benfey (1958).

William Bechtel and Robert Richardson (1993) pioneered the study of strategies for discovering mechanisms and emphasized the strategies of decomposition and localization in cases from the history of biology. Darden (2002; 2006, Section 12.2) discussed a subset of the reasoning strategies in this chapter. Philosophers and cognitive psychologists have extensively studied reasoning by analogy. Holyoak and Thagard (1995) analyzed the structure of analogical reasoning and listed what they consider to be the sixteen greatest analogies in the history of science. The philosophers Mary Hesse (1966) and Richard Boyd (1979) showed how analogies provide new theoretical terms. The cognitive psychologist, Dedre Gentner, studied humans doing analogical reasoning. She argued that good analogies are those that map relations rather than properties; see, e.g., Gentner (1983; 1997) and Gentner et al. (1997). Kevin Dunbar (1995) visited molecular biological laboratories to study reasoning in discovery. He found that they make greater use of "near analogies" to mechanisms in closely related areas within their own field, rather than "distant analogies" to more far flung analogues.

Of course, the list of types of mechanisms in Table 5.1 is very incomplete. The list is intended to give a flavor of the kinds of background knowledge about types available from contemporary biology, expanded by cases of currently rejected types from the history of biology. The ideas in this chapter build upon Norwood Russell Hanson's later view of retroduction, in which scientists infer from puzzling phenomena to a hypothesis of a certain *type*:

Schematically, [retroductive reasoning] can be set out thus:

(1) Some surprising, astonishing phenomena p_1, p_2, p_3 . . . are encountered.
(2) But phenomena p_1, p_2, p_3 . . . would not be surprising were *a hypothesis of H's type* to obtain. They would follow as a matter of course from something like H and would be explained by it.
(3) Therefore there is good reason for elaborating *a hypothesis of the type of* H; for proposing it as a possible hypothesis from whose assumption phenomena p_1, p_2, p_3 . . . might be explained (Hanson 1961, p. 630; italics added).

One must also add, as later critics made clear, (1) that H has not been ruled out on some other grounds (as discussed in Chapter 6) and (2) that H remains the most plausible hypothesis type after a systematic search of plausible schemas (see Josephson and Josephson 1994).

For more on T. H. Morgan's work, see Allen (1978) and Darden (1991).

For more on the discovery of reverse transcriptase, see Temin and Mizutani (1970), Baltimore (1970) and the discussion in Darden (2006, ch. 10). On forward chaining in the discovery of the mechanism of protein synthesis, see Crick (1958; 1988), Watson (1963; 1968), Watson and Crick (1953a; 1953b). On backward chaining by biochemists, see Hoagland (1955; 1959; 1996), Hoagland et al. (1959) and Zamecnik (1960). Zamecnik analogized their work to tunnel diggers (personal communication to LD). Their work is discussed in Rheinberger (1997), who emphasizes experimental systems rather than mechanisms, and Darden and Craver (2002), who focus on the discovery of the mechanism of protein synthesis.

Charles Darwin discusses the role of Malthus in his discovery of the theory of natural selection in his autobiography, published after his death in 1892 (Francis Darwin, ed.). An excellent two-volume biography of Darwin is by the historian Janet Browne (1995; 2002). The philosopher and historian of biology, Michael Ruse, investigated Darwin's early views about the analogy to artificial breeding; see Ruse (1973; 1979, pp. 170–71). Diane Paul (1988) discusses Spencer's phrase "survival of the fittest," and Darwin's adoption of it. Spencer used the phrase in his *Principles of Biology* (1864). Selection type theories in evolutionary biology, immunology, and the speculative theory of neural selection are analyzed in Darden and Cain (1989; reprinted in Darden 2006, ch. 8). The philosopher of science Mark Parascandola (1995) critically analyzes the NeoLamarckian view of Steele et al. (1998).

6 VIRTUES AND VICES OF MECHANISM SCHEMAS

INTRODUCTION

This chapter discusses criteria for evaluating mechanism schemas. First, we discuss a master list of virtues of scientific theories, virtues celebrated by the recent philosophy of science. Such virtues include explanatory power, simplicity, and coherence. One resounding lesson from the history of science is that such virtues are often sacrificed in scientific practice. None is an absolute criterion of good theories, though each is thought, in general, to be valuable. There are typically trade-offs among the different virtues in the search for mechanisms. Importantly, there are more specific virtues in the search for mechanisms: depth, completeness, and correctness. Phenomenal descriptions lack depth in that they describe a mechanism's overall behavior without describing the mechanism. Complete mechanism schemas have no significant black or gray boxes. Correct schemas accurately identify the components of the mechanism and show how these components are organized together. We discuss these specific virtues of mechanism schemas in the second half of the chapter.

VIRTUES OF GOOD THEORIES

Consider some virtues that philosophers and scientists alike describe as a basis for choosing one hypothesis or theory over its rivals. We divide these into four loose categories: formal features of the theory, pragmatic features (having to do with how useful the theory is), aesthetic features (having to do with the beauty and elegance of the theory), and empirical features (having to do with how well the theory fits the constraints that have been discovered through observation and experiment).

FORMAL VIRTUES

- *Testability*
  The theory should express commitments about the world that in principle can be confirmed or falsified on the basis of empirical evidence.
- *Internal Coherence*
  The theory should be coherent and should not contain contradictions.

- *Fertility*
  The theory should suggest new and exciting avenues of research. It should generate new research questions faster than it can answer them.
- *Conservatism*
  The theory should retain crucial bits of what came before and not break too quickly or too dramatically with tradition without a compelling reason for doing so.

AESTHETIC VIRTUES

- *Simplicity*
  The theory should posit only those entities, properties, causal relations, etc. that are necessary to account for the phenomenon.
- *Elegance*
  The theory should be compact and graceful.

EMPIRICAL VIRTUES

- *Empirical Adequacy*
  The theory should accommodate or fail to conflict with well-established phenomena.
- *Prediction*
  The theory should make accurate predictions and retrodictions, particularly concerning phenomena that would be surprising were the theory to be false.
- *Explanation*
  The theory should explain the phenomena in its domain either by showing how they follow from general laws of nature or by showing how they are produced, given rise to, or maintained by mechanisms.
- *External Coherence*
  The theory should be supported by (or at least consistent with) other well-accepted non-rival theories.
- *Generality*
  The theory should apply to more phenomena than its rivals.
- *Unification*
  The theory should unify diverse phenomena by showing them to be instances of a common pattern.

Of course, many of these virtues of theories are, in fact, also virtues of mechanism schemas. Mechanism schemas should be testable (testability) on the basis of empirical evidence when they are conjoined with background assumptions and auxiliary hypotheses. Mechanism schemas should not contain contradictions. Mechanism schemas should not be in conflict with well-confirmed scientific theories (external coherence) or well-established phenomena within the schema's domain (empirical adequacy) unless there is good reason for accepting the conflict. Mechanism schemas should, when instantiated, make predictions, and they should explain the phenomena in their domain, since all mechanisms are mechanisms of the things they explain. They should be supported by other relevant theories (external support). None of this is news. None of it is especially controversial. All of it must be hedged with qualification to accommodate great discoveries in the history of science that violated, or in some cases flouted, these apparent virtues.

In other words, the virtues of scientific hypotheses and theories are like other virtues: in making choices among hypotheses and theories, different virtues and vices must be weighed against one another. Sometimes one must break the rules to give birth to a new discovery or to nurture a nascent idea to maturity. Charles Darwin, for example, rightly stuck with his theory of evolution by natural selection even when evidence from physics appeared to contradict a fundamental assumption of his theory. The physicist Lord Kelvin used assumptions about the earth's cooling to estimate that the earth was much younger than Darwin had supposed. If the earth were as young as Kelvin proposed, the mechanism of natural selection would likely not have time to produce the observed abundance and diversity of species. The dispute about the age of the earth persisted for much of the last half of the nineteenth century. Because of this lack of external coherence between Darwin's theory and the physicist's estimates, Darwin made changes in later editions of the *Origin*. However, the problem persisted until the discovery of radioactivity in the early twentieth century. This discovery removed the problem about sources of the earth's heat and geologists extended the age of the earth to about 4.5 billion years. It turns out that there has been plenty of time for life to evolve. Of course biologists would be well advised not to build theories that openly flaunt physical constraints; but physicists, like biologists, can sometimes be wrong. Nascent theories would not survive if they were immediately rejected in the face of every apparently disconfirming piece of evidence.

Elegance and simplicity have clearly been important for many scientists. It is not at all clear, however, whether the notion of simplicity can gain much traction in the complex world in which biologists typically work. The evolved world is not as simple as it perhaps might have been. Many biological phenomena and mechanisms typically have myriad parts engaged in complex and dynamic interactions. Many are kludges, selected not for their simplicity or optimality but because they satisficed (were good enough) within available constraints of development and evolution. Some of the resulting mechanisms appear to be quite awkward and contrived relative to the designs of our best engineers. While it is true that one should not build gratuitously ornate mechanism schemas, there is considerable room for dispute about when the ornateness has become gratuitous.

We categorize simplicity here as an aesthetic virtue, although it might also be classed a pragmatic virtue. For testing purposes, it might be a good pragmatic strategy to start with a very simple hypothesis and then complicate it as the evidence demands. Again, simplicity typically must be weighed against other virtues, such as empirical adequacy and explanatory power, in the assessment of a theory.

Similar tradeoffs arise concerning generality and unification. Evolution by natural selection works by selecting variants that are adaptive in diverse environments. Hence, one should not be surprised to find theories in biology (other than natural selection) that are relatively limited in scope, applying only to a few species or to some types of cells. There might not be any unified theory that covers all or everything in a related domain of phenomena. As in the case of simplicity, where one wants the theory to be as simple as it can be under the circumstances, in biology we seem to demand that theories be as general or unifying as they can be without becoming (overly) vacuous or (overly) false.

In short, this traditional list of scientific virtues is at most a helpful starting place for thinking about how one ought to select among competing mechanism schemas. In the search for mechanisms, it is crucial to recognize that these virtues must be balanced against one another in the evaluation process. There are, however, more specific virtues and vices of theories that purport to describe mechanisms. So suppose that we start at a finer grain. Suppose we ask not, "What makes good theories good?" but rather, "What makes good mechanism schemas good?" Suddenly it is clear there is more to say. The instrumental goal of finding mechanisms places more specific constraints on the norms of theory selection than are required by these traditional criteria.

EVALUATING MECHANISM SCHEMAS: SUPERFICIALITY, INCOMPLETENESS, INCORRECTNESS

You can learn a lot about virtue by studying vice. So let us now ask: How might a mechanism schema fail? Evaluative strategies are designed to detect possible failures. There are three broad classes of failure: superficiality, incompleteness, and incorrectness. They are represented in Figure 6.1.

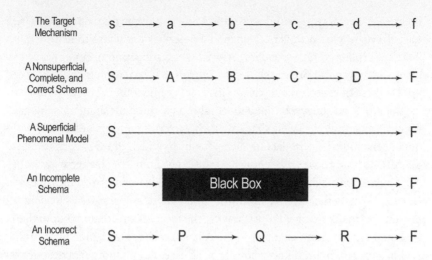

Figure 6.1 Superficial, incomplete, and incorrect schemas. S means start; F means finish.

SUPERFICIALITY

Superficial schemas, in the limit, merely redescribe a phenomenon. They do not describe, or purport to describe, underlying mechanisms. If they are correct, they may describe a correlation or a causal relation that can be used for prediction. However such schemas do not explain the phenomenon because they do not describe its mechanism (when the explanatory task is to find a mechanism, rather than, say, a mathematical law or a phylogenetic tree). They do not reveal the internal structure of the mechanism that produces, underlies, or maintains the phenomenon. They are accurately called phenomenal models.

The distinction between phenomenal models and mechanistic schemas is familiar in sciences outside of biology. In optics, Snell's law characterizes succinctly the relationship between the angle of incidence and the angle of refraction when light passes between two media. It does not explain why light refracts. To provide a mechanism schema, in contrast, one must appeal, for example, to facts about how light propagates or about the nature of electromagnetic waves.

Consider an example from cognitive science. Scott Kelso (b.1947) and colleagues have developed a precise mathematical description of a curious generalization about the coordination of limbs. If you wag your fingers back and forth in phase with one another, like the windshield wipers in most cars, and you gradually wiggle them faster and faster, you reach a point at which you can no longer hold the pattern. The fingers will suddenly shift so that they wiggle completely out of phase with one another. The phenomenon in this case is the shift from in-phase motion to out-of-phase motion. Kelso developed a model that describes precisely when this phase-shift occurs as a function of the frequency with which the subjects wag their fingers. This model describes the phase transition but it does not explain it. One wonders: Why does the transition occur at those frequencies? Why don't fingers shift from out-of-phase to in-phase as they speed up? On these questions the model is silent. It is a phenomenal description.

The difference between a phenomenal model and a mechanism schema is familiar to biologists as well. One might characterize patterns of heredity without understanding the molecular mechanisms by which those patterns are produced. One might have an accurate description of a phyletic lineage without understanding the mechanisms by which one species is transformed into another. One might know the ranges within which human blood pressure or plasma osmolality are maintained without knowing the mechanisms that maintain them. One might know that a change in a DNA base results in a change in an amino acid in a protein but not know how. In each case, the phenomenal description remains agnostic with respect to the mechanism by which the phenomenon is produced, underlain, or maintained. The model lacks the resources required to explain, rather than merely describe, the phenomenon.

Whether a phenomenal description is taken to be satisfactory depends upon the purposes for which the description is being used. If one wishes to predict when wagging fingers will transition from in-phase to out-of-phase movement, then a phenomenal description will suffice for such a purpose. However, when it comes to explanation and testing, the added detail in mechanism schemas is not merely gratuitous. Mechanism schemas are committed to more constraints about how the phenomenon is produced than are phenomenal models. By specifying the entities, activities, and organizational features that produce, underlie, or maintain the phenomenon, one undertakes more commitments about what the parts are, the effects of removing them, the results of experimental interventions, and the like. In this way, such mechanistic commitments suggest new experimental tests and observations. They also provide a basis for thinking methodically about how to intervene into a mechanism to control its behavior.

One extreme form of superficiality is embodied by *homuncular explanations*, which have appeared repeatedly in the history of biology. In a homuncular explanation, an ability or property of the whole is explained in terms of the same activity or property of a part. For example, homuncular theories of generation posited the existence of full-blown little men/women housed in the sperm or eggs. Ventricular theories of brain function attributed such faculties as cogitation and memory to distinct ventricles in the brain in an effort to explain how humans (wholes) cogitate and remember. Hans Driesch's (1867–1941) entelechies posit distinct life forces that explain the characteristic features of living organisms. Although homuncular explanations are not *ipso facto* illegitimate—there might have been little people in the sperm—they are rather limited in their explanatory force. At best, they localize an explanation to a particular part. At worst, they merely push the request for explanation one step back.

In sum, the distinguishing mark of a superficial, phenomenal model is that it describes the behavior of the mechanism without describing how the mechanism works. A schema explains rather than describes a phenomenon when the components in the schema correspond to components and organizational and productive features of the mechanism that produce, maintain, or underlie the phenomenon. Phenomenal models are superficial because they specify neither the internal components of the mechanism nor the organizational and productive features by which the mechanism works. Mechanistic models (instantiated schemas) have depth. They reveal the internal structure of a mechanism. This is why they are explanatory, not merely descriptive.

INCOMPLETENESS

Where phenomenal models do not even purport to reveal facts about mechanisms, incomplete schemas are best thought of as mechanism sketches. They have black boxes or question marks for components for which not even a functional role is known. In the nineteenth century, after the discovery that the link between generations of a species is provided by the germ cells (sperm and eggs), the question remained—how do the germ cells produce offspring that resemble their parents? In the twentieth century, genetics began to fill this black box with genes, DNA, and developmental mechanisms.

Sketches may also have gray boxes, where a functional role has been conjectured but there is no known occupier for that role. Darwin knew, for example, that his theory of evolution by natural selection required an account of the origin and inheritance of variations. Darwin and other biologists of the late nineteenth century speculated about theories of heredity, without much success. Darwin was fortunate that the problem of heredity could be cordoned off. For the purposes of Darwin's early theorizing, heritable variations could simply be documented to occur empirically (and Darwin had literally volumes of examples). At this stage of his discovery, he could afford to treat heredity as a form of replication across generations produced by some-mechanism-we-know-not-what. As long as heritable variations are available such that the better adapted ones can be selected, the mechanisms for producing variation and passing it between generations could be temporarily ignored. (Darwin wrestled throughout his career with what he called the "causes of variation.")

The goal in providing a complete description of a mechanism is to fill in black and gray boxes. Yet every description of a mechanism bottoms out at some point where the gain in detail makes no difference to the researcher. One can know a great deal about biochemical cascades in organisms, for example, without understanding much about quantum chemistry. Different scientists, with different exploratory, explanatory, and instrumental objectives, will often demand different degrees of detail about the mechanism in question. An evolutionary biologist might presume the very facts that physiologists are trying to explain. An electrophysiologist might count their understanding of a mechanism as complete when a protein chemist would see the explanation as full of holes. Although the ideal of completeness is relativized to particular scientific projects, with one scientist's black box counting as another's phenomenon-to-be-explained, once the required detail is specified, the deeper the mechanism schema goes into the hidden machinery by which things work, the more complete is the schema. We are aware of no mechanism schema in the history of science that is complete in this ideal sense. It is more appropriate, for this reason, to think rather in terms of degrees of completeness.

Here are three tests for the completeness of a proposed mechanism schema.

The "And How Does That Work?" Test
and the Vice of Boxology

This first test is a version of the familiar childhood game of iteratively asking "Why?," except in this case we ask, "And how does that work?" Let's take a familiar example: the eye. The retina transduces light from the environment into neural signals. *How does that work?* Perhaps you know that photons of light interact with photoreceptors and that this causes cells to become hyperpolarized, reducing their continuous release of neurotransmitters. *And how does that work?* It all begins when photons of light enter the cell and interact with rhodopsin molecules. *And how does this interaction work?* The photon isomerizes (changes the structure of) the molecule. *But that merely tells me that the molecule changes to an isomer form, but how does that work? How does light produce that change?*

You can play this game at any stage of a mechanism. When you can't provide an answer, you've found a gap in your understanding of the mechanism. Whether this gap matters for a particular project is a further question. Psychologists have shown that our everyday explanatory knowledge is really quite shallow for most things. Few people can pass this test for such mundane mechanisms of everyday life as those by which toasters and car engines work. We engage, in other words, in the everyday vice of *boxology*: input bread and electricity, push lever, output toast. Boxology is the vice of operating with incomplete schemas for which one cannot pass the "And how does that work?" test. But a goal of science is to push beyond the levels of understanding of everyday life to reveal the internal mechanisms by which things work. Mechanistic science is a corrective to our natural vices in this regard: it is a virtue of mechanism schemas that they allow one to answer iterative questions about how various parts of the mechanism work, and, conversely, how they fit into more inclusive mechanistic wholes.

The "What If That Worked Differently?" Test
and the Vice of Chainology

A second measure of the completeness (depth) of one's knowledge of a mechanism is the ability to answer questions about how the mechanism would behave if different parts of the mechanism were to work differently, or if they were to break, or if one were to add some new factor, such as a drug or a source of energy. Returning to the above example, one might ask: *What would happen if the rhodopsin molecule had a slightly different structure?* Or, to pick a simpler example: *What would happen to the transduction of light if one were to introduce a product that catabolizes rhodopsin?*

Scientists sometimes also indulge in *chainology*, boxology's twin. One becomes fascinated by nodes in a causal chain but loses sight of how the nodes work to produce, underlie, or maintain the phenomenon. One loses sight of the way that the individual components contribute to the behavior of the whole. To return to our optical example, one might entertain a chain:

Light → rhodopsin → transducin → phosphodiesterase → cyclic GMP → calcium channel → neurotransmitter

However, one might memorize this chain and still not understand how light is transduced in these photoreceptive cells. You know the links in the chain, but not how they all add up to phototransduction. Students with notes of this sort are not going to do very well on the test because they are not prepared to answer questions about how this mechanism would work under a wide variety of conditions. The students don't know the activities by which these nodes are linked together, nor do they know how these molecules are organized so as to know how changing one makes a difference to a subsequent one. They know very little about the resting state of the cell against which these actors stand out as making a difference. When students cannot say how the mechanism would behave if the links in the chain had been different, then their sketch is incomplete.

Compare this chainological description to the following, more detailed, mechanism schema: When light strikes a photoreceptor cell in the retina, rhodopsin molecules become photoexcited. The photoexcitation of rhodopsin activates a G protein, transducin, which in turn increases the activity of a phosphodiesterase. This leads to the hydrolysis of cyclic guanosine monophosphate (cGMP), closing cGMP-gated calcium ion channels in the plasma membrane. The resulting membrane hyperpolarization changes the rate at which the photoreceptor releases neurotransmitters onto downstream neurons. This is how photoreceptors convert light into neural signals. This description adds new details about how one node in the chain is linked to the next. It tells us, for example, whether one node is excitatory or inhibitory of the items at the next node, what the relevant properties of the various entities in the chain are, and it tells us the activity (e.g., hydrolysis, hyperpolarization) by which changes at one node give rise to the changes at the next.

Similarly, one might ask: What would happen if you used diesel gas in your non-diesel engine? What would happen if you used DC current on your toaster? If you cannot answer these questions, it is another example of the shallowness of your everyday understanding of these mechanisms. In this case, the shallowness reflects not a failure to fill black boxes with entities, but rather a failure to know what the relevant properties of these entities are and how those properties make a difference to the activities of the mechanism. Science corrects our everyday shallowness by digging deeper into the productive organization of the mechanism, thus ameliorating the vice of chainology.

The Build It Test

One sign of incompleteness is that one is unable to specify the schema in sufficient detail that one could use the schema as a blueprint for building a mechanism of that type. One might construct a mathematical model with variables corresponding to features of the mechanism and see if one can solve the equations. One might transform the schema into a program in a computer language and implement it on a machine. One might literally build a scale model according to the blueprint specified in the sketch. One might commandeer parts of a biological mechanism (such as DNA) to manipulate it for some practical objective. In each case, the ability to construct a working mechanism or a working simulation of the mechanism is an implicit test of completeness. Incomplete models won't work.

Our understanding of the mechanisms of light transduction, for example, has progressed to the point that scientists can begin to use their mechanism schemas as blueprints for engineering. On the basis of our understanding of photoreception in the retina, it is now possible to build artificial transducers that receive light through cameras and generate signals the visual system can use to distinguish different shapes and sizes, light and dark. At a deeper level, scientists' understanding of the family of opsin molecules of the sort described in the above chain has progressed to the point that they now can design them to work in a variety of ways (to excite cells, to inhibit them, to increase or decrease their excitability) and to control the rates and wavelengths at which they do all of these things. The now-emerging field of optogenetics (discussed in more detail in Chapter 11) uses these genetically modified opsins (taken originally from light-sensitive bacteria) to control the behavior of neurons by flashing light of different wavelengths through fiber optic cables.

Such maker's knowledge, increasingly evident across many areas of biology, is a mark of progress in our understanding of mechanisms. It is a mark of progress in our understanding of the basic mechanisms of genetics that we are now able to design genetically modified plants and animals with surprising accuracy and ease. It is a mark of progress in our understanding of the mechanisms of reproduction that we can now routinely create human life from sperm and egg cells under laboratory conditions and that we can clone sheep. It is a mark of progress in our understanding of HIV and its modes of transmission that has brought a raging AIDS epidemic under some measure of control. The practical utility of a mechanism schema for (a) allowing one to build a functioning mechanism from scratch, and (b) allowing one to modify the function of a mechanism at will are also epistemic virtues. We assess the accuracy and completeness of our schemas in part by the things that they allow us to accomplish.

There has long been a very close connection in the mechanical worldview between knowing how something works and knowing how to make it work. Knowledge-how is an indicator of the completeness of our knowledge-that. Yet it is an imperfect indicator. The practical limitations on human engineering and construction might prevent one from building the mechanism even if one knows all the entities, activities, and organizational features. Further, one might build a facsimile of a mechanism without understanding the target mechanism itself. Nonetheless, the power to reliably bring a mechanism under our control, or to build a functional equivalent from scratch, adds confidence that one's understanding of the mechanism is complete enough.

A complete description of a mechanism reveals its productive continuity; that is, it shows how each stage of the mechanism allows, prevents, or produces the next stage, without significant gaps, from beginning to end. One can evaluate the completeness of a schema by the reasoning strategies of asking how each stage works, by asking what would happen if it worked differently at some node, and by trying to build a replica from the blueprint in the schema. What counts as an acceptable degree of completeness varies from project to project, from one pragmatic context to another, but the degree of mechanistic completeness itself is not relativized; it is a plain fact whether more remains to be said about how the mechanism works.

INCORRECTNESS

Incorrect schemas and sketches fail to describe accurately the mechanism for the phenomenon. A mere how-possibly schema describes how the mechanism might work. A how-plausibly schema describes how a mechanism might work in a way that is consistent with the known evidence at a given time. A how-actually schema describes how the mechanism in fact works (or close enough for the purposes at hand). One goal in the discovery of a mechanism is to prune the space of how-possibly mechanisms to an ever narrower set of how-plausibly mechanisms and, when necessary, with hard work and good fortune, to a well-defined and unambiguous region of that space. The search process involves adding constraints to the space of possible mechanisms, showing that certain regions of the space can be ruled out or showing that the actual mechanism most likely lies in some region of the space of possibilities.

Correctness here should be understood in terms of a mapping between descriptive elements in one's mechanism schema and the entities, activities, and organizational features of the mechanism. If the schema is specified in terms of variables and mathematical relations, for example, the variables in the schema can be said to correspond more or less correctly to components, activities, properties, and organizational features of the target mechanism that produces, maintains, or underlies the phenomenon. The (perhaps mathematical) dependencies posited among these variables in the model can be said to correspond more or less correctly to the (perhaps quantifiable) activities among the components of the target mechanism. If the schema is specified in sentences, then there will be a mapping between elements in the description (the nouns, verbs, adjectives, and adverbs) and features of the mechanism. The difference between how-possibly, how-plausibly, and how-actually a mechanism works is a difference in correctness. Mechanists are garden-variety realists about such things: the goal is to describe correctly enough (to model or mirror more or less accurately) the relevant aspects of the mechanism under investigation.

As with completeness, correctness should be understood as a matter of degree. The acceptable degree of correctness varies from one project to another and one scientist to another. More importantly, however, many if not all mechanism schemas indulge in a certain amount of idealization about the mechanisms involved. Idealization is the deliberate falsity of a mechanism schema. One might understand the hyperpolarization of rod cells in the eye in terms of a model that treats the neuron like an electrical circuit that obeys Ohm's law, knowing full well that neurons technically do not obey Ohm's law. One might assume that a signaling molecule is more or less equally distributed in the cytoplasm for the purposes of constructing a mathematical model. As we noted above, criteria of empirical adequacy (here specified in terms of the correctness of a mechanism schema) often must be weighed against other virtues (such as the manageability or simplicity of a schema) in the choice of mechanism schemas. Here it is more appropriate to talk of a schema being correct enough rather than correct full stop.

CONCLUSION

Mechanism schemas, like other scientific theories, are evaluated in terms of a number of different virtues and vices that must be balanced against one another. In this chapter, we discussed traditional virtues attributed by scientists and philosophers of science to scientific theories, but noted that they are only a starting place for evaluating mechanism schemas. We discussed criteria for judging superficiality, incompleteness and incorrectness of mechanism schemas. And we noted vices to avoid, such as boxology and chainology. One indicator of a sufficiently complete understanding of a mechanism is the ability to build one, e.g., in a working computer simulation model or in an in vitro model of an in vivo mechanism made from off-the-shelf components. Another indicator is the ability to manipulate the phenomenon more or less at will by intervening on its component parts.

BIBLIOGRAPHIC DISCUSSION

On the general virtues of theories, see Newton-Smith (1981, ch. 9), Darden (1991, ch. 15). See also the extensive list of evaluative strategies in Figure 1 in Dietrich and Skipper (2007). On model testing, see Giere (1997), an elementary textbook on the subject. Elizabeth Lloyd (1987) discusses the difference between testing predictions of the model as a whole and testing claims about internal components of the model. For more on phenomenal versus explanatory mechanistic models, see Craver (2006), Kaplan and Craver (2011).

On Hans Driesch's view of entelechies, see Driesch (1929).

For Kelvin's views about the age of the earth, see Kelvin (1862a; 1862b; 1865) and Burchfield (1990). For details on the action of rhodopsin, see Hargrave (2001) and Kandel et al. (1990). On the Kelso (also called HKB) model of bimanual coordination, see Haken, Kelso, and Bunz (1985); see also Chemero and Silberstein (2008), Kaplan and Craver (2011).

Charles Darwin called his hereditary theory the "provisional hypothesis of pangenesis." He did not include it in his *Origin of Species* (1859). It was published in 1868 in the second volume of his *Variation of Animals and Plants Under Domestication*. Darwin's cousin, Francis Galton (1871), tested Darwin's pangenesis in rabbits; the results did not support Darwin's hypothesis. Darwin's pangenesis was one of many speculative theories of heredity in the late nineteenth century that were abandoned after the 1900 rediscovery of Mendel's work. For more on these theories, see Darden (1976; 1991), Dunn (1965) and Geison (1969).

Evidence for the shallowness and incorrectness of our everyday understanding of mechanisms is discussed in Wilson and Keil (1998).

Dretske (1994) makes the point that "If You Can't Make One, You Don't Know How It Works." Of course, not all mechanisms can be simulated, but being able to build a scale model or simulation is a good indicator of a complete how-possibly understanding. For further work on the relationship between making and knowing, see Craver (2010), Datteri (2009), and Datteri and Tamburrini (2007).

7 CONSTRAINTS ON MECHANISM SCHEMAS

INTRODUCTION: CONSTRAINTS AND CORRECTNESS

Mechanistic scientists place a premium on assessing whether or not a model or schema correctly describes a target mechanism. Correctness is a matter of degree of match, a correspondence between a mechanism schema and the target mechanism. How-possibly models describe a mechanism that might produce the phenomenon. How-actually-enough models more or less correctly describe the entities, activities, and organizational features of the mechanism. Commitment to the goal of correctness for mechanism schemas places a variety of empirical constraints on any acceptable mechanism schema.

The ascendancy of a mechanistic biology in the sixteenth and seventeenth centuries (that is, the rise of the unambiguous idea that biology is committed to the search for mechanisms) was paralleled by the ascendancy of a set of methods for vouchsafing the accuracy of one's claims about mechanisms. These methods provide the evidence by which mechanism schemas are evaluated. There is perhaps no more fitting demonstration of these features of the new science—mechanism and methodology—operating in parallel than William Harvey's (1578–1657) protracted empirical argument for the circulation of the blood. Although Harvey certainly retained many aspects of the Aristotelian worldview (including an emphasis on that-for-the-sake-of-which, or teleological, styles of argument and explanation), he nonetheless regularly used the term mechanism and its cognates, and he sought a coherent understanding or schema of how the parts involved in the movement of the blood are organized together to do something. Robert Boyle (1627–1691), René Descartes (1596–1650), and Thomas Hobbes (1586–1679), three central figures in the ascendancy of the mechanical philosophy during this period, all praise this aspect of Harvey's work. The historian, Richard Westfall, concurs with this assessment: "The essence of Harvey's demonstration of the circulation of the blood lay in his attention to the mechanical necessities of the vascular system" (Westfall 1971, p. 90).

More importantly for present purposes, Harvey also provides an exemplar of how to use a wide variety of kinds of empirical evidence to argue for a mechanism schema. In this chapter, we use Harvey's evidence against the then dominant Galenic view of the motion of the blood and in favor of his circulatory hypothesis to illustrate some of the kinds of evidence relevant to evaluating the correctness of a mechanism schema.

THEORETICAL BACKGROUND OF HARVEY'S DISCOVERY

In *De Motu Cordis et Sanguinis in Animalibus* (1628), translated *The Movement of the Heart and Blood in Animals* (1963), Harvey presents an extended series of arguments against the Galenic view of the motion of the blood and in favor of the hypothesis that the blood moves in a circle. According to Galen's theory (see Figure 7.2), blood manufactured in the liver is drawn out to the rest of the body by the attractive force of hungry tissues. Only a small portion of the manufactured blood makes its way to the heart, where the blood is heated. The innate heat of the heart causes it to actively dilate, drawing blood into its right chamber as "the dilation of a pair of bellows sucks in air, or as the flames of a lamp suck up oil" (Mowry 1985, p. 50). This dilation corresponds with the thumping of the heart against the chest. It also corresponds with the active dilation of arteries during the pulse. Once in the heart (shown in the more contemporary diagram in Figure 7.1), blood mixes with air from the lungs (delivered via the pulmonary vein to the left ventricle) and expels waste produced as the heart heats the blood. The blood then passes through minute (and invisible) holes in the septum into the left ventricle, where the blood is mixed with vital spirits. This blood is then distributed to the more tender organs of the body, such as the brain and the eyes, which require rarified nourishment.

Harvey's arguments against the Galenic view and in favor of his circulatory hypothesis build on the anatomical discoveries of his predecessors and teachers. Andreas Vesalius' (1514–1564) revisionist anatomical work fueled revolutionary attacks on Galenic anatomy just as Galileo's telescopic observations in the *Starry Messenger* (*Sidereus Nuncius*, 1610) rallied Copernicans against the Aristotelian/Ptolemaic picture of the solar system. Vesalius could find no pores in the septum of the heart despite repeated experiments to demonstrate their presence. Matteo Realdo Colombo (1516–1559) argued that the pulmonary arteries carry blood rather than air and waste. He posited the existence of the pulmonary ("lesser") circulation of the blood. And Fabricius ab Acquapendente (1533–1619), Harvey's teacher at Padua, had investigated the role of the valves in the veins. But it was Harvey who organized these disparate bits of evidence about the parts of this system into a substantially different kind of mechanism. According to his schema, "the blood is driven round a circuit with an unceasing, circular sort of movement" (87).

On Harvey's view (see Figure 7.3), the heart actively contracts and expels blood from its right ventricle into the pulmonary artery. The blood returns to the

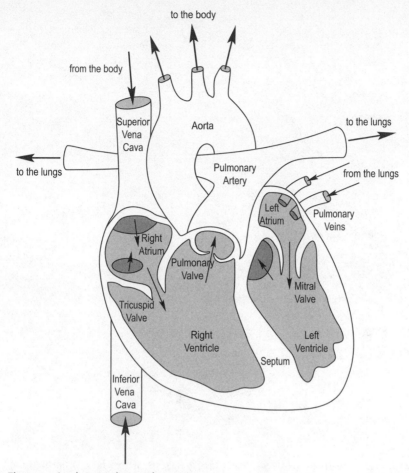

Figure 7.1 A schematic human heart.

left auricle via the pulmonary vein and is expelled again during the active contraction of the heart, whence it is distributed through arteries and to capillaries. The blood then returns to the heart through a system of veins. Although Harvey could not see any points of contact between the arteries and veins, he argued on the basis of evidence considered below that such a point of contact must exist. If so, the blood would complete a circuit through the body, just as water on the earth is converted into vapors and falls again as rain, and just as the planets complete their Copernican orbits about the sun (58–59).

These two competing schemas for the movement of the blood differ from one another most obviously in the kind of movement they attribute to the blood, that is, in their characterization of the phenomenon to be explained. For Galen,

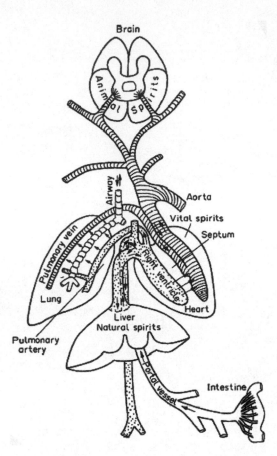

Figure 7.2
Galen's schema for the
movement of the blood
(Mowry 1985). Reprinted by
permission from Elsevier.

the blood moves from the liver to the hungry tissues, where it is consumed. For
Harvey, the blood circulates. It travels out from the heart to the lungs and the
rest of the body through the arteries, and it returns from them to the heart via
the veins. Galen and Harvey differ, just as significantly, in their understanding
of how the various parts of the mechanism underlying that phenomenon act
and are organized together. Much of Harvey's evidence for the circulation of the
blood turns on showing that his Galenic opponents cannot make sense of certain
facts about the active, spatial, and temporal organization of the components.
These facts constitute the evidential constraints on the space of possible mechanisms that rule out the Galenic schema as implausible and strongly suggest that
Harvey's schema is more correct.

Harvey considers a wide range of evidence in favor of his circulatory model
and against its Galenic competitor. These kinds of evidence are used to answer
different kinds of questions about the parts of the mechanism and how they are

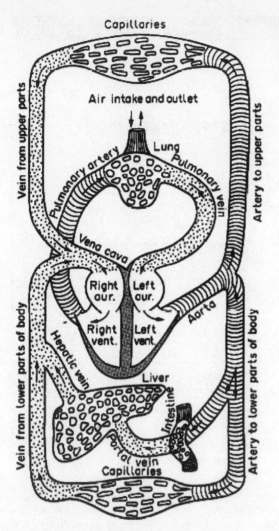

Figure 7.3
Harvey's schema for the movement of the blood (Mowry 1985). Reprinted by permission from Elsevier.

organized together in a whole. The kinds of evidence are distinguished from one another on the basis of the fact that they answer different questions that arise in the course of discovering a mechanism. Table 7.1 lists seven different kinds of questions that Harvey asked in the process of making his revolutionary discovery, each of which concerns a different kind of constraint on the space of possible mechanisms. Contemporary biologists continue to ask precisely these kinds of questions in their search for mechanisms for all sorts of phenomena.

The kinds of questions in Table 7.1 concern different empirical constraints on how the mechanism and its internal components can and cannot work, how

Questions about Evidential Constraints

Locations
Where are the components?
How are they arranged spatially with respect to one another?

Structures/Entities
What are the sizes, shapes, and orientations of the various entities?

Abilities
What can the component do/not do in the relevant circumstances?

Activities
What kind of activity operates at this stage of the mechanism?

Timing
What are the orders, rates, and durations of the various components in the
mechanism?

Roles
What is the functional role of the entity or activity in the context of the
mechanism?
What does it contribute to the behavior of the mechanism as a whole?

Production
What gave rise to or is given rise to by a given component in the mechanism?
How is the component situated within the productive chain of the mechanism?

Global Organization
What is the overall organization of the mechanism?
For example, does it contain forks and joins or feedback loops?

Table 7.1

they do and do not work, and how they are organized together into a mechanism. They guide the search through the space of possible mechanisms by eliminating some regions of that space and directing attention to others. We now discuss these questions, and the observations and experiments that Harvey and the anatomists of his day used to answer them, in more detail.

LOCATIONS

One central task in the search for mechanisms is to figure out where (in a molecule, cell, organ, body, or niche) different parts of a mechanism are, where they are located with respect to one another, and how they are connected together. Harvey made the discovery that the blood circulates by standing on the shoulders of anatomists who preceded him. For Harvey the primary goal was not so much to localize different components within the body; on this he mostly agreed with his Galenic opponents. He differed from them most significantly, however, in his understanding of how these parts act and in his understanding of the roles the different parts play.

Everyone agreed, for example, on the basic structure of the human heart: two atria, two ventricles. Everyone agreed that the right ventricle and left atrium connect to the lungs via vessels now known as the pulmonary artery and the pulmonary veins, though they ultimately disagreed about what these parts do. Everyone agreed that the vena cava enters the right atrium and that what we now call the aorta leaves from the left ventricle, though again they differed about the functions of these vessels. Nobody could begin to think about what these parts do without first understanding where they are located and how they are connected to one another. This fact is obvious but also extremely important for understanding the mechanical cast of mind.

For example, one crucial point in the debate concerned the means for getting blood into the left ventricle. According to the Galenic model, air rather than blood is delivered to the left side of the heart via the pulmonary veins (which also carry waste back to the lungs). Blood in the left ventricle, then, must pass through some other route, perhaps through invisible pores in the septum. Following Vesalius, Harvey will have none of this: "But, damme, there are no pores and it is not possible to show such" (19). Vesalius devised experiments to show that no such connection exists. He filled the right ventricle with water, and even with air, and then tried unsuccessfully to force the contents from one side to the other. Harvey's conclusion is that there is no point of communication between these two parts: they are not spatially connected to one another, so blood cannot flow from one to the other.

Harvey's discovery is fundamentally a physiological discovery about how the parts work together rather than an anatomical discovery about the locations of various parts. In many sciences, however, the search for mechanisms begins with the effort to localize components. In sciences dedicated to understanding cellular processes, it is often crucial to know whether an activity occurs in the nucleus or in the cytoplasm, whether an entity is membrane bound or whether it is free-floating. For example, the first step in the mechanism of protein synthesis (transcription) occurs in the nucleus and the second step (translation) occurs in the cytoplasm. Emphasis on localization is perhaps nowhere more evident than in the neurosciences, where a primary goal is to divide the brain into functionally distinct regions and to trace their anatomical connections with one another. As Harvey's work illustrates, anatomical facts about where items are and how they are connected to one another form a central spatial backdrop on which the behavior of a mechanism might be sketched. Knowledge of the locations of different entities and activities, in addition, guides researchers by telling them where to look to learn more about these entities and activities. In this way, facts about locations and connections place crucial constraints on the space of possible mechanisms.

STRUCTURES / ENTITIES

The structures of the entities in a mechanism can also provide crucial constraints on the space of possible mechanisms. By structure we mean such spatial features as the sizes, shapes, orientations, directions, and compartments of an entity, as well as material facts about its components and their structural features. Facts about the structure of an entity can be used to infer how the entity can and cannot behave and, correlatively, about its role in the system. Facts about structure also might provide clues as to how the entity itself was produced. In his discussion of the blood's movement, Harvey is more concerned with the former. In his discussion of development (in *De Generatione Animalium*), he is more concerned with the latter.

It is hard to complete a page of *De Motu* without encountering an inference grounded in details about anatomical structures. In his argument against the flow of blood through the septum of the heart, Harvey notes that the septum is "solid, hard, dense, and extremely compact," (19) and so unlikely to be a conduit for anything at all, let alone blood. In response to Galenic anatomists who distinguished the function of the left ventricle (carrying spirits) and the right ventricle (carrying blood), Harvey asks: "If the three tricuspid valves at the entry into the right ventricle hinder return of blood into the vena cava, and if the three semilunar ones at the opening into the artery-like vein have been made to hinder return of blood; how, when a similar arrangement holds in the left ventricle can we say that these valves have not been made similarly to hinder forward and backward movement of the blood?" He continues: "And as the valves are almost identically arranged in respect of size, form, and position in the left ventricle and in the right one, why do they say that they hinder the egress and regress of the spirits in the former, but of blood in the latter? The same sort of mechanism does not seem calculated to prevent equally effectively movements of both blood and spirits" (16). Harvey further appeals to the size of the heart, the thickness of its walls, and the relative volumes of the heart and the aorta (57–58). As we will see below, one of Harvey's crucial arguments is grounded in a careful study of the structure and orientation of the valves in the veins.

Again, it appears that facts about structure, broadly construed, constitute an evidential foundation for reasoning about how mechanisms work. Different kinds of mechanisms require different kinds of parts with different structural features. And in biological systems, the structures of those parts often provide clues about what they can and cannot do in the system. In this way, facts about the size, structure, and orientation of an entity often constrain the space of possible mechanisms in which it can participate.

A final example concerns the orientation of the semicircular valves spaced periodically along the length of the veins. On Galen's view, the blood flows out through the veins to the rest of the body. Harvey's teacher, Fabricius, performed dissections on veins and noted that their valves appeared to be oriented in such a way as to impede the flow of the blood away from the heart. Operating within the Galenic framework, Fabricius concluded that the valves must be in place to slow the blood as it moves away from the heart. On Harvey's view, in contrast, the blood flows back to the heart through the veins. The role of the valves is to prevent the blood from flowing in the opposite direction, back toward the periphery.

In his experiments on the hearts of a variety of organisms, Harvey noticed that valves come in pairs and that they appear to point toward the heart: "The twin valves (they occur mostly in pairs) at any one site face, and have contact with, one another; and in the extremities they are so ready to come together and act in unison that they completely prevent any backflow from the root of the veins into the branches, or from the larger into the smaller vessels. They are so placed so that the cornua of any one pair of valves face the middle portion of the sinuses of the next pair, and vice versa, all the way along the vein" (82). Even if blood could get past the first valve, in other words, there are numerous valves that it would have to contend with later. They are arranged, he says, "like the sluice gates which check the flow of streams" (83–84). The structures of the valves, in short, suggest that the blood in veins flows toward the heart and away from the periphery and not vice versa.

Robert Boyle describes Harvey as believing that this structural constraint, more than any other, suggested to him the very idea that the blood might circulate:

And I remember that when I asked our famous Harvey, in the only discourse I had with him, What were the things that induced him to think of a Circulation of the Blood, He answer'd me, that when he took notice of the Valves in the Veins of so many several Parts of the Body, were so plac'd that they gave free passage to the Blood Towards the Heart, but oppos'd the passage of the Venal Blood the Contrary way: He was invited to imagine, that so Provident a Cause as Nature had not so plac'd so many Valves without Design: and the Design seem'd more probable, than That, since the Blood could not well, because of the interposing Valves, be Sent by the Veins to the Limbs; it should be Sent through the Arteries, and Return through the Veins, whose Valves did not oppose its course that way. (Boyle V [1772] 1968, p. 427)

We return to this crucial example below. For now it simply illustrates the fact that learning the orientation of an entity in a mechanism often provides crucial insight into what that entity can do, clues as to where in the space of possible mechanisms one should look next if one is to pinpoint the target mechanism. Examples of this kind of inference are ubiquitous in contemporary molecular biology and biochemistry, where the conformations and orientations of molecules are often the crucial factors in allowing the entities to do what they do in a mechanism, in cytology, where cell structures are important determinants of their functions, and in evolutionary biology, in which structures such as beak size influence a creature's fitness.

ABILITIES

While facts about location and structure can provide clues as to what a given component can and cannot do (and so how a mechanism can and cannot work), often scientists test directly what a given component can and cannot do. Learning what the putative components of a mechanism can do, especially under the circumstances considered to be the normal operating conditions for the mechanism, constrains the space of plausible mechanisms for a phenomenon: the space of plausible mechanisms includes only mechanisms consistent with the abilities of the mechanism's components. Where an engineer must fashion components that have just the right properties to play their distinctive role in a mechanism (a wire conducts electricity, an amplifier magnifies a signal, and glass acts as an insulator), the biologist must investigate what the putative components of a mechanism are capable of doing as part of investigating what the component in fact contributes to the behavior of the mechanism. While this is a truism, it captures one of the central features of a test of a mechanism. As a strategy, one begins with some entity and perhaps some knowledge of its various features and structures, and one tests to see what the entity in fact can and cannot do.

Suppose that Harvey had stopped with his anatomical inspection of the valves in the veins. Grant that the structures and locations of the items strongly suggest that they impede the flow of blood away from the heart. It would still be open to Harvey's opponent to object that the blood behaves contrary to what one might expect on the basis of this inference. Indeed, Harvey's understanding of the anatomy of the valves was identical to that of his Galenic opponents, and they thought the valves merely slowed the blood on its way to the periphery. So Harvey tested the idea further. He tried to pass a probe (perhaps a glass tube) through the veins: "If I started from the root of these vessels and tried with all the skill I could muster to pass a probe in the direction of the small vessels, I was unable to do so over any great distance because of the obstacles provided by the valves" (83). Then he turned the vein around: "on the other hand, it was very easy to pass a probe from without inwards, that is, from the small branches toward the root of the veins" (83).

It was open to Harvey's Galenic opponents to object that blood is not a glass probe. Harvey's evidence, that is, is consistent with Fabricius' more conservative suggestion that the valves merely slow the blood. Furthermore, it is possible that the valves behave differently after the vein has been dissected than they do in the intact organism.

To rule out these alternative explanations and to make this conclusion, "more openly manifest" (84), Harvey conducted another experiment. Figure 7.4, the one and only diagram in De Motu, illustrates this experiment. You can do it yourself. It will take two people. First, use a belt or some other ligature (AA in Harvey's Figure 1, in Figure 7.4) to tie off the arm loosely above the elbow. Veins lie close to the skin. Arteries are deep. From Harvey's perspective (and ours), a loose ligature closes off the egress of blood through the veins while permitting blood to flow through the arteries. If you do it right, the veins in the arm will start to swell. (Technicians tie off your arm before they draw your blood because the veins are easier to find if they do.) As the veins swell, the valves become visible as small bumps or nodes in the veins (especially in "countryfolk," Harvey notes). These bumps are shown as B, C, D, D, E, and F in Harvey's Figure 1. (We do not know why there are two D's. Perhaps the engraver made an error?) Identify a node and press your finger on the vein just beneath it, that is, toward the hand. Pull the finger toward the hand along the vein (from O to H in Harvey's Figure 2). The blood above the node will not follow; it stops at the node. The blood will empty from the top portion of the vein and build up below the location of your finger. If you now try (as in Harvey's Figure 3) to push the blood from above the node into the space vacated when you earlier drew your finger along the vein (at O), you will not succeed.

To complete the story (Harvey's Figure 4 in Figure 7.4), let's start over with the ligated arm. Start by pressing on the vein a bit above the node (L). Use a second finger (M) to push the blood up the vein (away from the hand) and past the next valve (N). Notice that the blood passes with ease, but that the blood from above does not rush back to fill the vein vacated when you drew your finger upwards. Even in living creatures, blood is not able to flow backwards through these valves. The blood in the veins, he concludes, must move toward the heart.

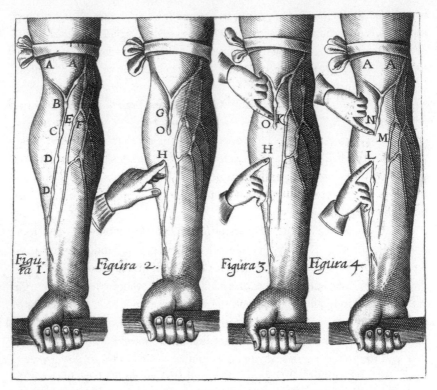

Figure 7.4 Harvey's experiment demonstrating how blood flows through the venous valves in the human arm (Harvey 1660). Reprinted by permission of Bernard Becker Medical Library, Washington University School of Medicine.

Given that the Galenic model predicts that the veins transport blood out to the body, Harvey's demonstration shows that the Galenic model is implausible. That model posits a role for the valves (slowing the blood on the way to the periphery) that they are not able to play. In fact, the valves prevent the blood from flowing in that direction. The veins, it would seem, do not have the ability to play the role of carrying the blood to the rest of the body. In contrast, Harvey's schema, in which blood departs the heart via arteries and returns via veins, is perfectly consistent with, is indeed supported by, this finding.

Ability constraints show what a component in a mechanism can and cannot do. Any how-actually schema must be consistent with findings about what the putative parts of the mechanism can and cannot do. Reasoning from ability constraints, like constraint-based reasoning generally, is a matter of testing the commitments of a given how-possibly schema. A schema, for example, posits that a component plays a given role in the mechanism. For the component to play that role in the mechanism, it is necessary that it have some set of abilities in the context in which it operates. If it has those abilities in those contexts, then the schema is consistent with what is known about the abilities of the part. If the part does not have those abilities, then the schema places unrealistic demands on the component, and so the schema is excluded from the space of plausible mechanisms. This form of reasoning can be found in any mechanistic science. A molecule that cannot bind to a given substrate cannot act as an enzyme for that substrate. A bone in the inner ear that cannot move freely cannot transmit vibrations in auditory mechanisms. A cell that cannot change its behavior in response to light cannot function as a phototransducer. Many scientists emphasize the importance of structure-to-function relations, but ability-to-function relations are just as important.

Findings about ability constraints thus tell the researcher what an entity can and cannot do. Typically, however, the abilities of a given entity far outrun those abilities by which the entity contributes to the mechanism at hand. Any given entity can do many things. Only some of the things it can do are part of the working of a mechanism.

ACTIVITIES

Ability experiments characterize the many things that a component can do. Suppose, however, that one knows that two items are causally related but one wants to know what kind of activity is involved. The goal is to look for telltale signatures of the activity as indicators. Different activities have different rates, durations, and energy demands. They act across different distances (such as the long-distance effects of hormones and the short-distance effects of electrical signaling in the nervous system, or the long-distance effects of gravitational attraction and the very short-distance effects of nuclear forces). They can work in one medium and not another. Experimenters can use these differences to infer which activities are relevant at a given stage of a mechanism.

Consider Harvey's disagreement with Galenic anatomists concerning the active motion of the heart. Galen thought that the activity (the active motion) of the heart is dilation. The heart expands and, in doing so, draws blood from the liver into its left atrium. When it expands, it thumps against the chest, creating the heartbeat. The veins and arteries also distend simultaneously, drawing the blood and vital spirits through them in like fashion. The veins, as it were, suck the blood out of the heart and distribute it. Harvey disagreed. He believed that the active motion of the heart (its activity) is contraction. Contraction forces blood out of the heart and through the arteries. The pulse, on his view (as ours) is created by the passive distension of the arteries and veins during the heart's contraction.

In search of evidence to decide among these hypotheses, Harvey dissected living animals to observe the heart working within the chest. He notes that the heart becomes harder (like a tense muscle) and more pale in color (because it is emptied of blood) as it decreases in size. He also notes that the fibers in the heart resemble those in muscles, which are known to contract forcibly when they are active. By analogy, he concludes that the heart is active in its contraction, not in its dilation: "in each movement it seems to rise up, gain in strength, diminish in size and harden, and its actual movement seems to resemble that of a muscle contracting in the line of its tendinous and fibrous components. Muscles in active movement gain in strength, contract, change from soft to hard, rise up and thicken; and similarly the heart" (27). Harvey thus hypothesizes that blood is actively forced out of the heart and into the arteries when the heart contracts and becomes smaller in size.

Harvey also criticizes the Galenic schema for the motion of the blood through the veins. How, he asks, do the arteries move the blood? If Galen holds that the arterial systole (contraction) draws the blood, "the impossible will come to pass, namely, arteries filling while they are contracting, or filling and not increasing in volume" (13). Furthermore, given that Galen held that the diastole (dilation) and the systole are everywhere simultaneous, it would not be possible for the blood to move as the result of such a mechanism: "For how can one of two so conjoined bodies, when they are simultaneously increasing in volume, draw from the other; or, when they are simultaneously contracting, receive from the other?" (13). Rather, Harvey notes, the blood always spurts from the end of a severed artery during the diastole of the artery, "never in the systole," indicating that "it is the force of the blood which causes the dilation of the artery" (14). How, he asks, could the artery so forcefully propel the blood as it is expanding? Harvey concludes that, "arteries increase in volume because they fill up like bags or leather bottles, and are not filled up because they increase in volume like bellows" (13).

These examples illustrate how learning about the activities of components in the mechanism places constraints on one's understanding of how the mechanism works. An adequate mechanism schema clearly cannot posit that a part acts in ways that it does not act. Nor can an adequate mechanistic schema omit an activity that makes a difference for the purpose at hand. Knowing the activities is crucial for knowing how mechanisms work, and learning the activities is often essential for pruning the space of possible mechanisms. Consider just a few other well-known examples. Otto Loewi (whose experiments we discuss in the next chapter) demonstrated that the vagus nerve influences the heart via chemical activities (the release of neurotransmitters) rather than electrical activities (such as the spread of a voltage change). The structural chemist Linus Pauling

(1901–1994) demonstrated that the activity of hydrogen bonding is crucial for the three dimensional structure of proteins. Alan Hodgkin (1914–1998) and Andrew Huxley (1917–2012) demonstrated that diffusion across a membrane could account for the form of the action potential in neurons. No mechanism has ever been discovered without a series of findings concerning the activities of its components.

Mechanisms are not mere chains of parts and properties; they have active parts. An adequate mechanism schema, one that avoids the vice of chainology, fills in the arrows of a sketch with the activities that constitute the productivity of the mechanism. If one knows the parts and properties of a mechanism but misunderstands the activities, one's mechanism schema is clearly inadequate. More importantly, however, knowing the activities in a mechanism can provide important clues about other parts of the mechanism. In the above example we see Harvey reasoning back and forth in a mechanism. If a given activity is to take place, then there must be certain parts with certain properties organized in a particular way. If a given activity has taken place, then one expects certain downstream consequences. Learning an activity thus allows one to backward chain to earlier prerequisites for the activity to occur and allows one to forward chain to the various signatures or marks that the activity will leave behind as a consequence of its action. In this way, activity constraints contribute to the discovery of mechanisms.

TIMING

Understanding a mechanism's behavior typically requires knowing not just which activities are involved, but also when the activities occur relative to one another. One might wonder whether a given component's activity is relevant at an early stage of the mechanism or at a late stage in the mechanism. One might wonder if the change in the strength of a synapse with repeated activity, for example, is the consequence of factors in the presynaptic cell (initiated when the neuron releases its neurotransmitters) or in the postsynaptic cell (in response to the effects of the neurotransmitter downstream). One might know that there are both plaques and tangles in the brains of patients with Alzheimer's disease but wonder which of these pathological signs appeared first.

Knowing the order, rate, and duration of the mechanism's stages constrains the space of possible mechanisms and also provides clues as to how the actual mechanism works. Most obviously, the arrows of Figure 7.3 guide us through successive stages in the flow of the blood through this circuit. There is a sequence of stages here from beginning to end, and it would not be possible to change their order without gumming up the works (or making it a different mechanism entirely). Similarly, the different stages have characteristic rates and durations that are crucial to the working of the mechanism, as is adequately attested to by the diagnostic value of the heart rate and blood pressure. Timing is everything for most mechanisms, and for this reason learning a mechanism's timing provides clues to how it works and to how it does not work.

Consider as a specific example Harvey's analysis of the motion of the heart. When one opens the chest of a living animal and observes the heart in action, the motions are so fast that it would appear to the naked eye as if everything happens at once. Harvey notes, "This is comparable with what happens in machines in which, with one wheel moving another, all seem to be moving at once. It also recalls that mechanical device fitted to firearms in which, on pressure to a trigger, a flint falls and strikes and advances the steel, a spark is evoked and falls upon the powder, the powder is fired and the flame leaps inside and spreads, and the ball flies out and enters the target; all these movements, because of their rapidity, seeming to happen at once as in the wink of any eye" (39). But as the animal begins to expire, the heart beats more and more slowly, and it becomes possible to see the separate motions of the heart independently. Harvey observes that the two atria contract simultaneously and then the two ventricles contract simultaneously: "There are, so to speak, two synchronized movements, one being that of the two auricles, and the other that of the two ventricles or the heart proper. These two movements are by no means simultaneous, but that of the auricles precedes and that of the heart [proper] follows, and the movement is seen to begin from the auricles and to pass on to the ventricles" (33). The movements of these different parts of the heart are also correlated with occurrences in the veins and arteries: "At the time when the heart is becoming tensed and contracting, and the chest is being struck, and in short systole is occurring, the arteries are being dilated, and producing a pulsation and are in their diastole. Similarly, at the time when the right ventricle is contracting and expelling its content of blood, the artery-like vein is pulsating and being dilated, synchronously with the other arteries of the body" (30).

It is not hard to follow Harvey's line of reasoning. If one wants to understand the operation of a machine, such as a clock or a gun, one needs to understand a sequence (or, at any rate, a temporal order) among the activities in the mechanism. First pressure is applied to a trigger. Then the flint falls. Then the spark flies. The heart is no different in this respect. The auricles contract together. Then the ventricles contract together. This contraction occurs with the arterial pulse. And so on. Harvey, in short, is decomposing the activity of the heart into distinct activities of its components and ordering those activities with respect to one another in time. Such ordering is crucial to figuring out how the mechanism works. A schema that posits an incorrect temporal order is inaccurate and so inadequate.

PRODUCTIVITY

Knowledge about the temporal order of stages in a mechanism is crucial in part because it indicates something about the flow of productivity in a mechanism. This is because producers precede their products in biological mechanisms. How does Harvey turn temporal facts about the movement of the heart's parts into facts about the productive sequence of the mechanism? In this case Harvey provides an answer through an experimental manipulation, which, in Harvey's words, "is more illuminating than the light of noon" (68).

Harvey invites the reader to cut into a live snake, because the movements of the heart are slower and more visibly distinct in snakes than they are in mammals. He notes that if you clamp the vena cava, perhaps with your thumb and forefinger, "you will see at once an almost complete emptying, through the pulsation, of the part between the fingers and the heart, the blood being drawn out of it by that pulsation" (68–69). The heart becomes lighter in color and smaller in size, even in its distended state, and eventually it appears to die. The effect is reversed as soon as one releases the clamp. If, in contrast, one clamps the arteries leaving the heart, one can see the arteries begin to bulge. The heart swells and turns bright purple to the point that you "believe its suffocation to be immanent." Releasing the clamp, however, returns everything to its normal color and size. The path of the blood goes from the veins, through the heart, and out through the arteries.

This experiment goes beyond facts about location and timing to reveal a productive sequence in the mechanism. It shows that blood enters the heart and is expelled by its forceful contraction. No amount of detail about the relative timing of the contractions in the heart and the ventricles alone would entail that the blood moves in one direction rather than the other. But by showing how interventions early in the mechanism (before the heart) influence what comes afterward (the blanching, the shrinking), Harvey shows how the movement of the blood in the arteries is produced. The heart shrinks and blanches because no blood flows into it. By showing how interventions late in the mechanism (after the heart) impact upon the function of earlier components (expanding the heart, turning it purple), Harvey is able to assemble these disparate temporal parts into a productive sequence.

Mechanisms are productive. They are how things work from beginning to end. An adequate schema of a mechanism reveals that productive order. Learning which stages of a mechanism create, drive, maintain, manufacture, or otherwise produce which others is thus essential to learning about a mechanism. And facts about this productive order are thus crucial constraints on the space of possible mechanisms and essential guides to the overall working of a mechanism.

ROLES

Another crucial part of discovering a mechanism is learning the roles of the mechanism's components. A role (or function) is the contribution that the component makes to the mechanistic context in which it operates (see Chapter 2). The question is not "What can this component do?" as in ability constraints, or "How does this part act?" (as in activity constraints), but more narrowly, "What does this part in fact do that makes a difference to the mechanism's behavior?" More colloquially, one might ask: "What function does the component serve within the context of the mechanism?" Reasoning from role constraints is straightforward. A possible schema describes a set of components playing different roles in the mechanism. These components are postulated to accept different inputs, to give different outputs, to perform different operations. By learning the role of a component, one can then focus attention on the region of the space of possible mechanisms containing schemas in which the component plays that role.

We have already seen that Harvey and Fabricius disagreed about the roles of the valves. Fabricius thought that they slowed the blood, and Harvey demonstrated that they could not play that role. More significantly, however, Harvey and Galen disagree about the very function of the heart. Indeed, the only time Harvey uses the word role is when he discusses the function of the heart. In Harvey's fifth chapter, titled "The movement and functional activity of the heart," Harvey collects the above findings about the temporal sequences of the heart's motion and its relationship to the movements in the veins and arteries, to draw a conclusion about the role of the heart's movement. He argues by analogy with the mechanisms of swallowing: the root of the tongue rises against the top of the mouth, driving water into the fauces, the larynx is closed, the top of the gullet opens, and then different muscles push and draw the fluid down toward the stomach. In like fashion, the different parts of the heart muscle allow it to "swallow" the blood from the veins and push it into the arteries: "When a horse drinks and swallows water, one can see that the swallowing and the passage onwards of the water into the stomach occur with successive gullet movements, each one causing a sound and an audible and tangible thrill. In similar fashion, with each of those heart movements there is a transmission of a portion of blood from the veins into the arteries, and during it the occurrence of a pulse which is audible within the chest" (40). Thus he concludes:

. . . the heart's one role is the transmission of the blood and its propulsion, by means of the arteries, to the extremities everywhere. Hence the pulse which we feel in the arteries is nothing but the inthrust of blood into them from the heart. (40–41)

In this argument, Harvey relies both on facts about the temporal sequences of the heart's motion and on an analogy with the known mechanisms of swallowing to draw conclusions about how the motion of the heart fits into the mechanism for the movement of the blood: it does not suck the blood in from the veins. Rather, it expels the blood into the arteries.

The vice of chainology, in which one knows the components of a mechanism without really understanding how they contribute to the behavior of the mechanism as a whole, is avoided in part by thinking functionally about different components. If one understands how a mechanism works, one should be able to say something about the role that the different components play in the operation of the mechanism. One should know, for example, what would happen to the operation of the mechanism were the part in question to behave differently or not at all. One should be able to say which features of the component are and are not relevant to the behavior of the mechanism. The teleological form of thought embodied in reasoning about an item's role, a form of thought that asks what the different components and activities are for, or what they contribute to a mechanism, is not merely a *façon de parler* maintained as yet another vestige of Aristotelianism in Harvey's thinking. Rather, it is crucial to understanding how biological mechanisms work: to seeing how a part fits into the organization of the mechanism as a whole.

GLOBAL ORGANIZATION

The last kind of constraint on mechanism schemas we shall discuss concerns the overall organization of a mechanism. In particular, we are here concerned with the overall productive structure of a mechanism. Does it work sequentially, from beginning to end? Does it contain forks and joins? Is it organized rather as a cycle? Clearly, the centerpiece of Harvey's schema is a feature of the overall organization of the mechanism: that the blood flows in a complete circle: out through the arteries to the body and back via the veins. Harvey acknowledges that he is not in a position to complete the circle. He does not know how the blood from the arteries makes its way into the veins. Yet he is certain that the blood must somehow complete that task.

To see why Harvey was so convinced, we need to consider his master argument, a semiquantitative argument that spans five of the seventeen chapters of

De Motu. Harvey describes this argument as "so novel and hitherto unmentioned" that he not only fears he will "suffer the ill-will of a few, but dread[s] lest all men turn against me" (57). The argument is only semiquantitative because it involves no precise measurements and no actual calculations (so far as one can tell from the text). One might consider his master argument a kind of thought experiment if it were not fundamentally constrained by facts about the spatial and temporal features of the mechanism in question.

Harvey asks the reader to estimate with him the volume of the left ventricle (a feature of the mechanism's spatial organization) and, from that, the volume of blood that (as he has shown above) is expelled from the heart with each contraction. Here, Harvey is not entirely precise: "let us feel that we may, by a reasonable inference, declare the amount ejected into the artery to be a quarter or a fifth or a sixth, or at least an eighth, part of the dilated ventricles content" (62). In humans, Harvey estimates it to be about half an ounce (or three drachms), or at least a third that much.

Harvey then asks the reader to consider the rate at which the heart beats. Again, the precise values do not matter so much. In half an hour, he estimates, we typically find over a thousand beats, though in some people we might find as many as three or four thousand. Here is a widely accepted temporal constraint on the mechanism.

Now Harvey puts them together, multiplying the ejected volume by the number of heartbeats. Whatever reasonable value one chooses to accept for the volume of expelled blood and for the rate of the heart, Harvey notes, the amount of blood expelled by the left ventricle in the course of half an hour will be greater "than can be found in the whole body." Precisely the same calculation will also allow one to estimate the amount of blood passing through the pulmonary circulation from the right ventricle. As a result he concludes, "the beat of the heart is continuously driving through that organ more blood than the ingested food can supply, or all the veins together at any given time contain" (63). There is not enough food consumed to supply all the new blood that would be required (contra the Galenic view), and there is far more blood than a body can hold, unless, that is, the blood circulates (in favor of Harvey's view). Thus do quantitative facts about the spatial and temporal organization of the mechanism combine to reveal something about the overall organization of the mechanism: that the blood moves in a circle rather than sequentially, from start to finish.

How-possibly experiments of this sort are found throughout the sciences. Scientists build models to describe phenomena, and, in many cases, they can specify their model with sufficient precision that they can make precise predictions as to how the target mechanism will behave under a variety of circumstances. The behavior of those models can then be compared to the behavior of the target mechanism to see if the model could possibly explain the phenomenon. Galen's model fails this test. Harvey's in contrast passes. Despite the fact that Harvey could not complete the circle, showing how the blood from the arteries makes its way into the veins, he was thus convinced that the whole mechanism must be organized in a circle. Otherwise, there would be more blood than the body could make or store.

CONCLUSION

Harvey's arguments for the circulation of the blood provide a powerful illustration of the kinds of questions that scientists ask as they search for mechanisms. These questions concern various constraints on the working of the mechanism. These constraints are the kinds of evidence that shape the space of possible mechanisms. By learning facts about the abilities, activities, locations, roles, structures, and temporal features of components and stages in a mechanism, one prunes from the space of possible mechanisms those schemas that are inconsistent with the constraints. One also provides instructive clues about where, in that space of possible mechanisms, the correct mechanism is likely to be found. In the next chapter, we take a closer look at specifically experimental constraints and a variety of ingenious experimental designs deployed in the search for mechanisms.

BIBLIOGRAPHIC DISCUSSION

Galen of Pergamon (129–216 AD) was a prominent ancient physician and philosopher whose medical views were widely influential up to the time of the scientific revolution. For selections from his works see the Singer translation, Galen (1997).

In this chapter we work directly with the 1963 edition of Franklin's translation of Harvey's *De Motu Cordis* of 1628, published as *The Circulation of the Blood and Other Writings*. All page references to Harvey are from that work. Mowry (1985) provides an accessible introduction to the dispute between Harvey and Galen. Whitteridge (1971; 1977) provides a more detailed look at Harvey's intellectual development and his debt to other anatomists, particularly Colombo. Against the tendency to emphasize Harvey's Aristotelian and teleological commitments, Whitteridge argues that Harvey is "concerned with the understanding of how various phenomena happen, not why they do" (1971, p. 128).

For more on the relationship between Harvey, Aristotelianism, and the mechanical philosophy, see Bylebyl (1973; 1977; 1982), French (1994), and Westfall (1971). Readers might consider viewing the short film, *William Harvey and the*

Circulation of the Blood (Welcome Film Collection), which discusses Harvey's work in context and demonstrates many of the experiments described in this text. For a discussion of Harvey's comparative methodology, a feature of Harvey's writing we have not emphasized, see Lennox (2006a; 2006b). For more on the role of constraints in reasoning about mechanisms, see Craver and Darden (2001), Darden (2006), and Craver (2007, chs. 4 and 7).

8 EXPERIMENTS AND THE SEARCH FOR MECHANISMS

INTRODUCTION

Experiments are indispensable in the search for mechanisms. To experiment on a mechanism, in the sense with which we shall be concerned, is to place one or more parts of the mechanism under the experimenter's control and to assess from its responses to such interventions how the mechanism must be organized and what its working parts must be. Mechanism schemas express, implicitly, a variety of commitments about the results of such experiments. This is why experimental results can be used to prune and shape the space of possible mechanisms. Different kinds of experiments are used to ask different kinds of questions about how a mechanism works. Our goal in this chapter is to discuss these questions and the experimental strategies used to answer them. We show how these experiments contribute to the construction, evaluation, and revision of mechanism schemas.

Specifically, we discuss three loosely distinct kinds of experiments. The first kind tests whether one entity, property, activity, or organizational feature is causally relevant to another. For example, one might seek the relevant background conditions and triggers for a given phenomenon, or assess whether one putative stage in the mechanism is in fact relevant to what occurs at a later stage. The second kind of experiment, componency experiments, tests whether a given entity, property, activity, or organizational feature is relevant to the behavior of a mechanism as a whole. Such experiments are used to assess relationships between levels in a hierarchically organized mechanism. The third kind of experiment involves patterns of intervention and detection arranged to address particular questions about a mechanism. This third kind of experiment, which encompasses the bulk of the experimental work in mechanistic sciences, is more heterogeneous than the first two. Our goal in this chapter is not to offer a systematic taxonomy of experimental types but rather to call attention to the ways that experiments of the first two sorts are frequently combined, either in a single experiment or across multiple experiments, to answer specific questions about how a mechanism works.

EXPERIMENTS FOR TESTING CAUSAL RELEVANCE

The entities and activities at one stage in a mechanism produce, stimulate, maintain, inhibit, or prevent the entities and activities at another stage. In order to do this, those in the earlier stage must be capable of making a difference to what happens in later stages. Experiments that test for causal relevance test whether a given entity, property, activity, or organizational feature makes a difference to what happens at a later stage in the mechanism.

One can recognize a number of different kinds of difference makers within this broad class. Some difference makers are triggering conditions, conditions that start the mechanism working as a fly's touch triggers the macabre clamping mechanism in Dionaea muscipula (the Venus flytrap). Some difference makers are background conditions, such as the presence in the trap of a store of fluids in cells and energy gained through photosynthesis. Some difference makers are remote, such as the origins of the clamping mechanism in the species' evolutionary past or in the development of a particular plant. Other difference makers are more proximal, such as the interaction between the fly's body and the trigger hairs. Still others are further downstream, such as the difference that moving a trigger hair makes to the spreading of an electrical potential across the two sides of the trap. Some difference makers produce or augment a later stage in the mechanism, as the movement of hair cells produces an action potential across the plant lobes. Other difference makers are inhibitory, preventing or inhibiting something downstream (as when the plant is locked, loaded, and waiting for prey). In this section we discuss experiments for testing difference making, or causal relevance, in many of its guises.

EXPERIMENTAL SYSTEM, INTERVENE, DETECT

Consider an uncontroversial example of a test of causal relevance from the history of epidemiology: Joseph Goldberger's work in the etiology of pellagra. Pellagra is a disease characterized by dermatitis, diarrhea, dementia, and ultimately death. The dermatitis resulting from pellagra is so severe that it was commonly mistaken for leprosy. The disease was first identified in eighteenth-century Spain, and it ravaged the American South during the late nineteenth and early twentieth centuries. Between 1907 and 1940, approximately three million Americans contracted pellagra, and approximately thirty thousand of them died. In 1914 Joseph Goldberger (1874–1929) was commissioned by the U.S. Attorney General to head an investigation into the causes of this disease.

The dominant idea at the time was that pellagra is an infectious disease, spread by germs passed through contact and contaminated foodstuffs. Goldberger's epidemiology data from prisons and asylums, where pellagra was unusually common, made him doubt this hypothesis. The disease was common among inmates but extremely rare among prison guards and officials, and Goldberger thought it implausible that germs could so thoroughly discriminate against the inmates. Noting that inmates in the South were commonly fed a corn-heavy diet, and relying upon earlier hypotheses that perhaps the disorder was due to a toxin in spoiled corn, Goldberger began to entertain the hypothesis that the disorder was dietary in nature. It is in part due to Goldberger's efforts that we now know that pellagra results from a lack of niacin or tryptophan in the diet.

To test his dietary hypothesis, Goldberger first decided to change the corn-centered menu at two Mississippi orphanages and at the Georgia State Sanitarium, places where pellagra was particularly problematic. He substantially increased the volume of meat and eggs in the diet. The results were dramatic: those suffering from pellagra recovered, and nobody eating the varied and fresh diet contracted the disease.

To test whether a corn-based diet could in fact have been the cause of the disorder, Goldberger arranged to feed eleven healthy prisoners from Mississippi's Rankin State Prison Farm a steady diet of corn. (The Governor of Mississippi offered to release them from prison in exchange for their services.) Six of the eleven developed pellagra.

Goldberger's idea that dietary deficiency causes the disease was poorly received by those in the South, who took the suggestion as an affront to their dietary practices. Continued acceptance of the germ hypothesis frustrated Goldberger's efforts to convince people to institute the dietary reforms that his experiments suggested were necessary. In a heroic effort to overcome this resistance, Goldberger attempted to infect himself, his wife, and his assistant with pellagra using all the known means of infection. He injected himself and his volunteers with blood from pellagra patients. He swabbed the noses and throats of pellagra patients and then wiped them on the noses and throats of his healthy volunteers. He placed the scabs from the skin of pellagra patients, and at times bits of feces, into capsules and fed them to his volunteers. Although many of them felt queasy after ingesting the capsules (understandably), none of the volunteers contracted pellagra.

Goldberger's experiments argued convincingly first, that pellagra was not transmitted by germs; second, that he could induce pellagra with a diet restricted to corn; and third, that pellagra could be cured and prevented by a healthy and varied diet. Although he was not in a position at that time to say precisely how diet made a difference to whether or not a person succumbs to pellagra, Goldberger was in a position to claim that contact with other pellagra victims did

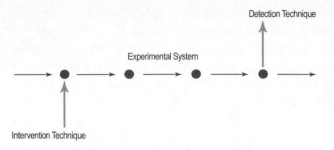

Figure 8.1 An abstract experiment for testing causal relevance.

not appear to be even causally relevant to the disease and that, instead, a diet restricted to corn was positively relevant to the disease.

This colorful example illustrates some of the main features of an experiment for testing whether one entity, property, or activity is causally relevant to another. An abstract representation of a standard experiment of this sort is shown in Figure 8.1. A given causal or mechanistic schema is instantiated in a given *experimental system*. One *intervenes* on the experimental system to change one or more putative cause variables, and one *detects* to see the effect, if any, of that intervention.

The *experimental system* (i) is the organism, system, or sample group on which one performs the experiment. Goldberger's experimental systems were orphans, prisoners, and healthy volunteers from his family and laboratory. He chose his subjects as a more or less representative sample of a larger class (human beings). As it is no longer considered ethically acceptable to knowingly expose human beings (let alone children and prisoners) to the possible causes of a potentially deadly disease, such experiments are routinely done in nonhuman animal models or in isolated systems, such as organ systems or cells grown in a culture. Indeed, as Goldberger focused on determining precisely what about the diet made the difference to the occurrence of pellagra, Goldberger began using dogs as animal models, as certain strains of dogs exhibit a black tongue disease, like pellagra, that also results from amino acid deficiencies.

The second component of this classic picture is the *intervention technique* (ii). The intervention technique is a means of introducing a change into one or more of the features in the mechanism. In controlled experimental manipulations, one literally sets the value of the intervention variable to the desired value. In other cases, one might find natural circumstances under which the intervention variable will be set to that value. In either case, one must consistently be aware of whether the intervention technique changes only the variable of interest or whether it changes in addition other variables in the mechanism that might account for the difference (if any) observed between experimental and control groups. In Goldberger's first two experiments, he intervened on diet: in the first instance, replacing a predominantly corn diet with a mixed diet, and in the second, replacing a mixed diet with a diet predominantly of corn. In his third experiment, Goldberger relied upon background knowledge about how germs are passed from one person to another to intervene in a way that would transfer such germs if they exist: injecting blood, transferring mucosal secretions, and consuming scabs and fecal matter.

Four standard conditions on interventions in this canonical treatment of experiments are represented graphically in Figure 8.2. Unidirectional arrows represent causal relations, bidirectional dotted arrows represent correlations, and bars across arrows indicate that the relation is absent. The goal is to intervene so as to find out whether X makes a difference to Y. In this figure, an intervention (I) changes the value of X, removing other causal influences, U, on X (as in I4). This intervention produces a change in Y that is not mediated directly (as in I1), or by affecting an intermediate variable S (as in I2), or by being correlated with some other variable C that can change the value of Y (as in I3). Control groups often receive sham interventions so that the researcher can ascertain whether the difference in the effect variable is due to the change in the target variable or due to some other, unintended, consequence of the intervention.

We say these conditions are idealized because experimenters often lack means for intervening in ways that satisfy these constraints or, alternatively, because they are not in a position to show that all of these constraints are, in fact, satisfied. Goldberger's interventions, in fact, were relatively imprecise; he intervened on many variables at once. For example, Goldberger not only increased the portion of meat and eggs in the diets of his subjects, he also decreased their consumption of corn. As a result, he could not firmly conclude that the addi-

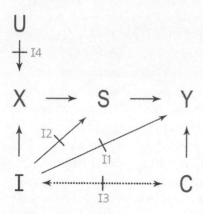

Figure 8.2
Idealized conditions on interventions in a test of whether X causes Y. Arrows represent causal relations. Crossed arrows indicate that the indicated causal relationship should not hold. Dotted arrow represents a correlation.

tion of meat, rather than the removal of corn, explained the disappearance of the pellagra symptoms. In the language of Figure 8.2, one wants to contrive an experimental setup in which any observed change in Y (pellagra incidence) can be attributed to a change in X (e.g., amino acids in the diet). However, because Goldberger's experiment violated condition I1 (the intervention itself introduced many changes that might produce the observed effect directly rather than via an amino acid deficiency), he could conclude only that something about the dietary change appeared to make a difference. The later experiments on black tongue in dogs allowed researchers to more carefully control the diets and other preventative interventions, allowing them ultimately to identify niacin as the primary dietary difference maker. Similarly, an experiment that changed the living conditions of the experimental subjects relative to the population from which they are drawn (e.g., moving the pellagra victims to a sanitized ward) would foil the experiment along path I3. For this reason Goldberger instructed his collaborating institutions not to change the sanitary conditions for the children, patients, or inmates during the experiment as some measure of control. Despite their limitations, Goldberger's experiments were sufficient to flag diet as potentially causally relevant and to cast serious doubt on the germ hypothesis for the etiology of pellagra. They showed that something about the diet likely makes a difference to the incidence of the disease, whereas exposure to known vectors for the transmission of germs did not make such a difference.

The third component of this classic view is the *detection technique* (iii), a device, indicator, or other process that takes some feature or magnitude in the experimental system as an input and returns a reading or measure as an output. This detection technique registers the difference, if any, between the experimental group and the control group. Goldberger simply monitored the presence of absence of symptoms in his experimental population. A reliable detection technique is one that indicates well: it is tightly correlated with the phenomenon one is trying to detect in such a way that one can say, with greater or less precision, how the feature or magnitude in question has changed as a result of the intervention. Litmus paper, for example, reliably turns from blue to red in an acidic solution (pH less than 7). The extent of an induced muscle twitch can be used to indicate neurotransmitter release from a motor neuron. Sometimes indicators are connected to their targets in very complicated ways. Functional imaging of the brain, for example, relies upon a complex chain: changes in neural activity (specifically, dendritic field potentials) change the oxygenation of hemoglobin in the blood, which then changes the rate at which hydrogen ions recover their spin after being placed in a magnetic field, which alters the radio signal they generate. That signal must then be detected, processed, fit to brain maps, and statistically analyzed before researchers can draw any meaningful conclusions about brain activity. What matters is not the number of steps in a detection technique, but rather the reliability of the indicator: how well changes in the indicator reflect changes in the feature or magnitude of interest.

Of course, controlled experimentation is an important method for testing hypothesized mechanism schemas. However, to design an experiment that rigorously tests a claim about the active organization of the mechanism, one often has to know or presuppose a great deal about what the parts of the mechanism are likely to be and how they are likely to be (and not be) organized. Meaningful experimentation (with useful interventions and detections) can take place only against a wealth of background knowledge about the active organization of the system under study. In his third experiment, for example, Goldberger relied on known means for transmitting diseases. In his first two, as we noted above, Goldberger's imprecise interventions ran roughshod over a number of potential causes of pellagra. A seemingly good experiment can be foiled by a failure to take into account unintended consequences of one's interventions or artifacts for one's detection techniques.

Experiments of this sort provide evidence about what makes a difference to what is in a mechanism. Such experiments are used to determine the start or set-up conditions for a mechanism and to establish the active organization among its components, that is, to establish how something happening at one stage of the mechanism produces, inhibits, modulates, or otherwise makes a difference to what happens at other stages.

INTERLEVEL EXPERIMENTS FOR TESTING COMPONENTIAL RELEVANCE

Experiments are also used to assess interlevel relationships in a multilevel mechanism. Interlevel experiments are used to answer two related questions. First, one might ask whether a given part is relevant to a given explanandum phenomenon at a higher level, that is, to the behavior of a mechanism as a whole. Second, starting with a higher-level phenomenon, one might wonder which parts in the system are relevant to the phenomenon (that is, which parts are components in the mechanism for that phenomenon). The purpose of interlevel experiments is to determine which parts are relevant to the behavior of a mechanism as a whole. As such, interlevel experiments play a crucial role in integrating findings at multiple levels into a single mechanism schema in which more and more gray boxes can be turned to glass boxes.

As shown in Figure 8.3, interlevel experiments can be bottom-up or top-down. On the left is a bottom-up experiment, in which one intervenes into a component in a mechanism and detects changes in the behavior of the mechanism

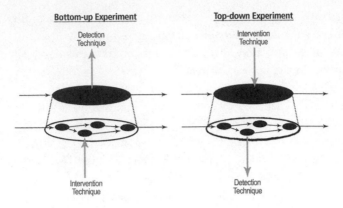

Figure 8.3 Interlevel experiments.

as a whole. On the right is a top-down experiment, in which one intervenes to manipulate the phenomenon and detects changes in the activities or properties of the components in the mechanism.

To see how such experiments work, and how they address the above questions, we need to look more closely at the three most common kinds of interlevel experiments: interference experiments, stimulation experiments, and activation experiments. These kinds of experiments differ from one another depending on whether the experiment is top-down or bottom-up, and depending on whether the intervention is excitatory or inhibitory.

INTERFERENCE EXPERIMENTS

Interference experiments are bottom-up, inhibitory experiments. In interference experiments, one intervenes to diminish, disable, or destroy some putative component in a lower-level mechanism and then detects the results of this intervention for the explanandum phenomenon. If the component is in fact relevant to the phenomenon, then removing the component or otherwise inhibiting its behavior should make some difference to the phenomenon. If the component plays an excitatory role in the mechanism, then removing the part should prevent or in some way impede the phenomenon. If the component plays an inhibitory or regulating role, then removing the part should augment, excite, or at any rate, change the phenomenon. To claim that some part is a working part or component is to commit one's self to the fact that the part makes a difference, and if the part makes a difference, then interfering with the part should have consequences for the mechanism's behavior. If the part does not make a difference to (is not relevant to) the phenomenon in question, then inhibition of the component should be of no consequence. (Such experiments might be foiled by redundant parts or by compensatory responses in the system in question, two possibilities that always must be kept in mind when dealing with biological systems. Such experiments might also be foiled if one intervenes on a generic background condition for the mechanism in question. However, here we put these issues aside to focus on the structure of the experimental inference in broad outline.)

Interference experiments are common in all mechanistic sciences. Case studies in clinical neuropsychology, for example, investigate how people's cognitive abilities are impaired by diseases, strokes, or accidental damage to brain regions. Lesion studies in neuroscience play the same role: one removes a region of the brain to see how the rest of the brain functions in its absence, and one tries to infer from the observed deficits just what (if anything) the given part contributed to the behavior of the mechanism. Gene knockout studies are often used to remove functional proteins from a system in order to assess whether the given protein in fact makes a difference to the phenomenon. In other cases, one might use designer pharmacology or toxins of various sorts to selectively block or impede putative components in a mechanism. Claude Bernard (1813–1878), often referred to as the father of experimental medicine, made a career out of inhibitory experiments: removing organs in order to see just which phenomena are affected by that intervention.

STIMULATION EXPERIMENTS

Stimulation experiments are bottom-up, excitatory experiments. In stimulation experiments, one intervenes to excite or intensify some putative component in a mechanism and then detects the effects (if any) of that intervention on the explanandum phenomenon. As with inhibitory experiments, the assumption is that if a given part is a component in the mechanism for a given phenomenon, than one should be able to influence the phenomenon by stimulating the part. If the part is an excitatory component in the mechanism, then stimulating the part should augment, induce, or otherwise alter the phenomenon in question. If the part plays an inhibitory role in the mechanism (such as an inhibitory neurotransmitter), then stimulating it should inhibit the phenomenon. If, in contrast, the part plays no role in the mechanism, then stimulating the part should have no consequences for the behavior of the phenomenon. (Here we assume that the intervention technique in question targets only the part in question and does not spread to other parts, violating the above assumptions on intervention techniques. We also assume that the mechanism in question does not compensate for the added load delivered to one of its parts. Again, we are interested here in the general structure of the inference.)

Like inhibitory experiments, stimulation experiments are ubiquitous in biology and other mechanistic sciences. In neuroscience, one might intervene to stimulate a part by delivering an electrical stimulus to a brain region, perhaps by using designer pharmacology or toxins to excite a given population of cells. In genetics, one might alter a regulatory element, leading to the constitutive production of a given protein. In physiology, one might contrive any number of means to drive a given component. In community ecology, one might do a pulse release (of organisms) that increases the density of one species in a community and then measure the effects on the community as a whole. But in each case the general structure of the experiment is bottom-up and excitatory.

ACTIVATION EXPERIMENTS

The last kind of interlevel experiment is the activation experiment. In activation experiments, one intervenes on the start conditions to activate, trigger, or produce the explanandum phenomenon, and one detects the behavior of one or more putative components of its mechanism to see if they change as a consequence. These excitatory, top-down experiments are represented on the right side of Figure 8.3. The basic assumption behind activation experiments is that if a part is a component in the mechanism for a given phenomenon, then the part should change in some way depending on whether the phenomenon is or is not manifest. In the most intuitive case, the part would become active, or would increase its activity from baseline, when the phenomenon occurs. However, if the component plays an inhibitory role in the mechanism, then one might in fact see the part decrease its activity from baseline during performance of the task. If the component plays no role in the mechanism for a given phenomenon, then (all things equal) one should not expect the component to change during the operation of the mechanism. (Such experimental methods might be foiled, for example, if the part in question is an inconsequential by-product of the working of the mechanism or if the part need not change its behavior when the mechanism is engaged, as a flywheel on a piece of machinery continues to turn even when the energy input mechanism is not engaged. Again, we neglect such possibilities to focus on the general structure of this kind of experiment.)

Examples of activation experiments are easy to generate. In fPET and fMRI studies in neuroscience, one engages a subject in a cognitive task while monitoring the brain for markers of activity, such as blood flow or changes in oxygenation. In single and multiunit electrophysiological recording experiments, one engages the subject in a task while recording the electrical activity in neurons. In some experiments, researchers monitor the production of proteins, or the activation of immediate early genes such as c-fos and c-jun, during the performance of a task in order to tell whether a given population of cells in fact contributes to the performance of that task. In an example we consider in detail below, Julius Axelrod watched how radiolabeled neurotransmitters move in and out of neurons as they fire.

Like experiments for testing causal relevance, interlevel experiments involve intervening and detecting. The difference lies fundamentally in the number and location of the interventions. For example, in an interference experiment, one often intervenes to remove a part *and* to activate the mechanism as a whole, for it is only under such intervention conditions that one can test whether the part plays a role in the output (or overall behavior) of the mechanism. (For example, one might lesion the hippocampus in a rat and then run the rat in a maze to see if the hippocampus makes a difference to the ability to run mazes.) In stimulation experiments one typically intervenes in one of the components in order to detect its consequences for the output of the mechanism as a whole. In activation experiments one might intervene to elicit the behavior of the mechanism as a whole in order to detect changes in one or more of the component parts. Interlevel experiments, therefore, are not different in kind from experiments for testing causal relevance, but rather should be viewed as particular kinds of such experiments. Different patterns of intervention and detection, in other words, answer different questions about whether and how a given part contributes to the behavior of the mechanism as a whole.

COMPLEX EXPERIMENTS FOR ASKING
SPECIFIC MECHANISTIC QUESTIONS

Although the above kinds of experiments (causal and interlevel) tend to be the primary focus of abstract discussions of experiment, many experiments in biology fail to fit this intervene-and-detect structure exactly. We close this chapter by considering some alternative kinds of experiments that do not fit neatly into this standard perspective: by-what-activity experiments, experiments with multiple interventions, series of experiments to reveal organization, and experiments relying on contrived experimental systems.

BY-WHAT-ACTIVITY EXPERIMENTS

Suppose that one has a mechanism schema with a gray box, but that one is unable to say what kind of activity fills the gray box. That is, suppose that one knows the inputs to and outputs from the gray box, and that one has established by experiment that the inputs are causally relevant to the output, but that one does not know the kind of activity by which the output is generated from the input. How might one use experiments to solve this problem?

Otto Loewi (1873–1961) faced exactly this kind of problem in his early-twentieth-century research on sympathetic (specifically vagal) control over the heart. Loewi knew that the heart is connected to the vagus nerve. He also knew that if one intervenes to stimulate the vagus nerve, the heart slows its beating and beats less intensely. Stimulating another nerve, one that Loewi called the accelerator nerve, speeds up the heartbeat and makes the heart beat more intensely. But Loewi did not know the activity by which these stimuli make a difference to the behavior of the heart.

Two alternatives presented themselves. According to the chemical hypothesis, the nerves communicate with the heart (and other organs) by releasing chemical substances, now known as neurotransmitters. According to the electrical hypothesis, in contrast, nerves communicate with the heart through the flow of electrical current: the nerve literally sparks the organ into action or, as the case may be, into submission. Loewi struggled to devise an experiment that would distinguish these activities, but eventually it came to him, as if in a dream (at least if we trust Loewi's retrospective report):

> The night before Easter Sunday of that year I awoke, turned on the light, and jotted down a few notes on a tiny slip of think paper. Then I fell asleep again. It occurred to me at six o'clock in the morning that during the night I had written down something most important, but I was unable to decipher the scrawl. The next night, at three o'clock, the idea returned. It was the design of an experiment to determine whether or not the hypothesis of chemical transmission that I had uttered seventeen years ago was correct. I got up immediately, went to the laboratory, and performed a single experiment on a frog heart according to the nocturnal design. (Valenstein 2005, p. 58)

Here is one version of his experiment: Loewi isolated two frog hearts, one with the nerves attached, and the other with nerves removed. He put each heart in its own bath of Ringer's solution (an isotonic salt solution). He used an electrode to stimulate the vagus nerve connected to the first heart. As expected, the heart slowed down. Then he used a pipette to transfer the Ringer's solution from the first heart's bath to the second heart's bath. The second heart slowed. After washing the hearts and allowing them to return to their baseline activities, he repeated the experiment, this time stimulating the accelerator nerve. As expected, the first heart began to beat more rapidly and intensely. As one can now predict, when he transferred the solution from the first heart's bath to the second heart's bath, the second heart sped up as well. When he washed out the Ringer's solution, the heart went back to its baseline beating.

Loewi's experiment is designed in such a way that only chemical transmission, not electrical transmission, would allow the effect to be transmitted via a pipette from one heart to the other. The molecular transmitters would, he reasoned, diffuse into the Ringer's solution and could thus be carried by pipette from one container to the other. An electrical mechanism, in contrast, would not continue to work under these conditions. Although electrical signals do spread through Ringer's solution, they do not accumulate in the fluid and would not be carried along in the pipette. The ability to transfer the effect from one heart to the other in this manner strongly suggests that the chemical hypothesis is correct: the nerve can influence the heart by releasing a chemical substance. Loewi thereby placed the concept of a chemical neurotransmitter, a concept now taken as fundamental to our understanding of the nervous system, on firm experimental footing.

Loewi's experiment is a *by-what-activity* experiment. The goal is to discover which kind of activity fills the gray box between a cause and its effect. Loewi's late-night design involved contriving a mechanism (the two communicating hearts) that would work if one kind of activity filled the gray box and that would not work if another kind of activity did the job. Loewi's experiment commandeers a crucial difference in the mode of action by which these activities work. Unlike a chemical signal, an electrical signal would not accumulate in the bath and could not be poured from one bath to the next. Loewi's result thus provides compelling evidence that the mechanism involves chemical transmission as a crucial activity.

Although we have coined the term *by-what-activity* experiments, such experiments are common throughout biology. One might, for example, wonder whether the mysterious activity inside a gray box is affected by changes in temperature (as thermodynamic activities are), or whether it depends on the presence of ATP, or whether it can be preserved in the absence of various sorts of pharmacological antagonists, or whether it works in the medium of a particular cellular fraction (separated out with a centrifuge), or whether it requires protein synthesis. In short, one finds a precondition for the occurrence of a given kind of activity, and one compares situations in which that precondition is met with cases in which it is not. So long as the different kinds of activities under consideration do not rely upon the same preconditions, such experiments can then be used to discriminate among competing hypotheses about what activity is in a given gray box. That, in a nutshell, is the structure of a by-what-activity experiment.

BY-WHAT-ENTITY EXPERIMENTS

Suppose that one has a mechanism schema with a gray box but that one is unable to say what kind of entity fills the gray box. That is, suppose that one knows the inputs to and outputs from the mechanism and that one has established by experiment that the inputs are causally relevant to the output, but that one does not know the working entity in the gray box. How might one use experiments to solve this problem?

Oswald Avery (1877–1955) faced this kind of problem in his study of bacterial transformation, a process by which non-virulent bacteria become virulent. He and his colleagues' discovered that DNA, not proteins, is the key working entity in this process. This finding is now viewed as one of the first hints that DNA is a key working entity in genetics. Consider the experiment by which Avery established this exciting result.

The bacteria that cause pneumonia, *Pneumococcus*, can be either virulent (active in producing the disease) or avirulent (and thus innocuous). Before Avery's work, researchers showed that mice injected with a small amount of avirulent, so-called Type II bacteria, and a large amount of heat-killed Type III bacteria (virulent, if not killed) died from a Type III infection. They hypothesized that something in heat-killed Type III bacteria had *transformed* (their term) the living Type II bacteria into the virulent Type III. But they did not know which entity in the heat-killed extract had the transforming activity.

Avery and his colleagues outlined the goal of their research: "The major interest has centered in attempts to isolate the active principle from crude extracts and to identify if possible its chemical nature or at least to characterize it sufficiently to place it in a general group of known chemical substances" (Avery, MacLeod, and McCarty 1944, p. 175). Their goal was to find the key working entity in the mechanism of bacterial transformation. They noted that transformation had never been observed to occur spontaneously. They developed an experimental system, which they called the "reaction system," so that they could reliably produce transformations *in vitro*. They found that even a carefully filtered extract of Type III bacteria transforms living, avirulent Type II bacteria into Type III. Furthermore, they found that the newly transformed bacteria reliably produce virulent, Type III offspring. The change, in short, was heritable.

They then set about to purify and characterize the active chemical substance in the extract. By varying the chemical substrates they put into the experimental system, and by using success in bacterial transformation as an assay for whether the working parts of the mechanism were in place, they were able to investigate which chemical substrates are and are not involved in the transformation mechanism.

They found that agents that destroy RNA (ribonuclease) and protein (trypsin, chymotrypsin) fail to prevent bacterial transformation. In contrast an enzyme then called desoxyribonuclease, which destroys DNA, does prevent transformation.

Robust evidence poured in as researchers used many other experimental techniques, all of which implicated DNA as the culprit. They had good evidence for the chemical nature of the working entity, but they concluded: "In the present state of knowledge any interpretation of the mechanism involved in transformation must of necessity be purely theoretical" (Avery et al. 1944, p. 189). They had filled a gray box with an entity, but other stages of the mechanism between injection of DNA and the production of the transformed type of living bacterium remained a black box.

Avery enumerated some of the how-plausibly hypotheses about the type of mechanism: maybe the DNA should be "likened to a gene"; perhaps it is a virus, such as the possible agent in fowl tumors (later shown to be an RNA virus that reverse transcribes its RNA into DNA); perhaps, the ever-cautious Avery allowed, some minute amount of some other substance survived the available purification methods and is so intimately associated with the DNA as to escape detection (that is, perhaps the intervention technique changes too much at once). Avery concluded: "If the results of the present study on the chemical nature of the transforming principle are confirmed, then nucleic acids must be regarded as possessing biological specificity the chemical basis of which is yet undetermined" (Avery et al. 1944, p. 191). That is exactly what happened. As a result of Avery's work, the biochemist Erwin Chargaff (1902–2002) used new techniques in 1950 to find that DNA is a more complex molecule than biochemists of the 1930s had reported. DNA, one of the constituents of chromosomes, became a more viable candidate for the entity identified with genes.

By-what-entity experiments work much the same way as by-what-activity experiments. One develops an experimental system in which the phenomenon of interest can be induced. In this case, Avery needed to induce bacterial transformation. To determine the entity involved in such a chemical mechanism, one might attempt to see which purified chemical products are necessary for the phenomenon to be induced or maintained. One might find a way to eliminate or disable an entity and see if such an intervention prevents the phenomenon from being induced under the standard set-up conditions. Alternatively, one might successively compile different mixtures of entities to see which components, when combined, suffice to make the mechanism work. This, in abstract, is the structure of a by-what-entity experiment.

SERIES OF EXPERIMENTS AND MULTIPLE INTERVENTIONS
As discussed above, Otto Loewi's experiments placed the concept of a neurotransmitter on solid experimental footing in physiology. Later researchers extended this idea, showing that it applied not only to the vagus nerve and the heart,

but also to most synapses in the nervous system (although some synapses are, in fact, electrical). Others worked to isolate and characterize the ever-growing list of chemical substances that function as neurotransmitters in the brain. One precondition for the continued effectiveness of chemical neurotransmission between neurons, or between neurons and organs such as the heart, is that there be some means to clear the neurotransmitter from the synapse. Otherwise, neurotransmitters would build up and continue to act on the postsynaptic neuron or organ even when the neuron is not active.

Although Loewi did not know it at the time he performed his experiment, the neurotransmitter connecting the vagus nerve to the heart is acetylcholine (ACh). ACh, first described by Henry Dale (1875–1968) in 1914, is also the dominant neurotransmitter responsible for communication between motor neurons and muscles. When ACh binds to receptors in the muscle, the muscle contracts. If ACh were allowed to build up in the synapse between them, the muscle would continue to contract even when the contract signal is not being sent down the neuron. For this reason these synapses contain an enzyme, known as acetylcholinesterase, which rapidly degrades the released neurotransmitter and prevents it from building up in the synapse. Acetylcholinesterases are found in snake venoms, poisons, and weaponized nerve agents. They can cause complete paralysis when all the muscles innervated by ACh contract forcefully. As one would predict from Loewi's experiments, they can also cause the heart to slow and stop. Such enzymes are also used to treat certain neuromuscular disorders, such as myasthenia gravis, by effectively increasing and prolonging the effect of ACh on the postsynaptic muscles.

Acetylcholine is just one of over one hundred neurotransmitters in the nervous system. A major research project for neuroscience after Loewi's and Dale's pioneering work has been to understand how different neurotransmitters are synthesized and inactivated. This research has opened new pathways for experimental intervention into the brain and, more importantly, new tools to correct brain malfunction.

Between 1958 and 1961, Julius Axelrod (1912–2004) carried out a series of experiments focused on the inactivation of the neurotransmitter norepinephrine (also known as noradrenalin) at synapses in the sympathetic nervous system. The sympathetic nervous system is a portion of the autonomic nervous system that regulates many basic bodily homeostatic functions (such as heart rate and blood pressure) and that triggers fight-or-flight responses when an animal is under stress. Two effects of norepinephrine are especially relevant below: increased blood pressure and the control of salivation. Axelrod wanted to know how norepinephrine is inactivated.

When Axelrod began his work to discover the mechanism for this phenomenon, it was a reasonable hypothesis that all neurotransmitters are inactivated by the same sort of enzymatic transformation that inactivates ACh. Axelrod's series of experiments disconfirmed that hypothesis and revealed a new kind of mechanism for regulating neurotransmitters. The main steps in his research were as follows.

In the first stage, Axelrod identified an enzyme in rat liver that inactivates norepinephrine. This research involved homogenizing rat livers (in a blender), exposing the homogenate to norepinephrine, and measuring the likely products of norepinephrine metabolism. In particular he found that something in the liver inactivates norepinephrine by replacing a hydrogen atom with a methyl group (which has the chemical formula CH_3). He identified the enzyme responsible for this reaction, and called it catechol-O-methyltransferase (COMT). He then hypothesized that perhaps COMT inactivates norepinephrine in the sympathetic nervous system as well.

In the second experiment, Axelrod used two interventions to test this hypothesis. First, he injected rats with norepinephrine to increase their blood pressure. Then he injected them with known inhibitors of COMT. He expected that this second intervention would prolong the effects of norepinephrine on blood pressure. It did so, but only to a small extent. This suggested to Axelrod that the synapses responsible for the sympathetic control of blood pressure must use some form of deactivation in addition to COMT.

To investigate further, Axelrod used some radiolabeled norepinephrine. He and his collaborators developed sensitive assays for detecting the presence of these molecules and then used tracers to track the location and by-products of norepinephrine metabolism. He injected cats with large doses of radiolabeled norepinephrine, and he found that the chemical tends to concentrate in organs richly innervated by the sympathetic nervous system (such as the salivary glands and tear glands). More surprisingly, he found that the radiolabeled molecules in those structures were not metabolites of norepinephrine, but rather norepinephrine itself; it was not being metabolized. So he conjectured that perhaps the sympathetic nervous system regulates norepinephrine, not through chemical inactivation, but rather through a kind of reuptake. Perhaps, he reasoned, sympathetic nerves remove the norepinephrine from the synapse, sequestering it inside the cell. This reuptake mechanism is illustrated in figure 8.4.

To test this reuptake hypothesis, he and Georg Hertting (b. 1925) made unilateral lesions to the superior cervical ganglia of cats, killing the nerves innervating the eyes and the salivary glands on one side of the body. In a second intervention they then injected the cats with labeled norepinephrine. They found that the

Radiolabeled Norepinephrine Concentrates in Live Axon Terminals

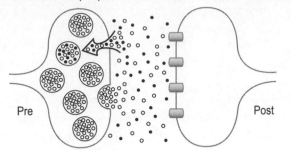

Figure 8.4
Axelrod's experiments
demonstrating
norepinephrine
reuptake. Dark dots
represent radiolabeled
neurotransmitters.

Stimulated Axon Releases Radiolabeled Norepinephrine

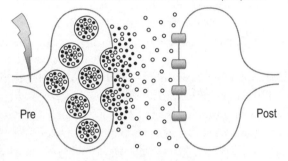

Cocaine Blocks Reuptake of Norepinephrine

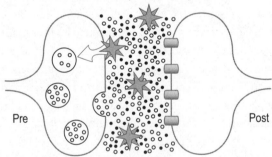

labeled neurotransmitter concentrates on the intact (innervated) side, and not on the lesioned side; the dead neurons do not take up the norepinephrine. In a third intervention, they then stimulated the live sympathetic nerve and showed that the sequestered, labeled transmitter is in fact released from the neurons. In a fourth intervention they showed that they could prevent the labeled transmitter from being re-sequestered by treating the nerves with cocaine. (It is now known that cocaine produces its prized psychoactive effects by blocking the reuptake of neurotransmitters and enhancing the effect of endogenous neurotransmitters.) Research quickly revealed that other transmitters, such as dopamine and serotonin, are also taken up into nerve terminals and that their reuptake can also be regulated with drugs.

Taken together, these experiments provided lasting insights into the mechanisms of the chemical synapse. These findings have led to the design and selection of drugs to control the central and peripheral nervous systems. By the end of the millennium, drugs that inhibit serotonin reuptake, such as fluoxetine (brand name, Prozac), were routinely used to treat depression, and they continue to be used today.

Two points are worth emphasizing. First, experiments for discovering new mechanisms often involve multiple interventions into one and the same experimental system. Clever experiments in biology frequently involve combining multiple forms of intervention and detection at different loci in a mechanism or set of mechanisms. Axelrod, for example, intervened by injecting radiolabeled norepinephrine, killing neurons, stimulating the live sympathetic nerves, and observing, at different stages, the location of the radiolabeled norepinephrine. Second, series of experiments (taken together with spatial and temporal constraints) can be arranged cumulatively to point convincingly to a new mechanism schema or to a particular region in the space of possible mechanisms. While experiments for testing causal relevance and interlevel relevance are crucial in the search for mechanisms (because they place evidential constraints on the active organization of the mechanism), one does not truly understand experimentation until one understands how multiple different interventions and detections are combined to answer different kinds of questions about the structure of a mechanism. Different stages of such series of experiments are used to prune the space of possible mechanisms of empirically inadequate schemas. Other stages of experimentation are used to assess directly one or more types of schemas for a given phenomenon.

PREPARING THE EXPERIMENTAL SYSTEM

We end with one further example that emphasizes the results of the first three, but adds another component that is ubiquitous in contemporary biology: explicit preparation of the experimental system. Most research in laboratory biology is carried out with biological systems that have been prepared and manipulated to serve specific scientific purposes (to facilitate inference, to make intervention easier, to assuage ethical concerns, and to increase one's degree of control over the system). Sometimes the system is prepared by selective breeding to produce genetically identical organisms. Sometimes the system is prepared by removing it from its organismic context. Sometimes it is maintained in stable background conditions (as Loewi did for frog hearts with Ringer's solution). The use of model systems is an indispensable part of biological practice, not merely because it has become the norm but because the logic of experimentation in contemporary biology often demands that the experimental system be isolated and prepared so that the experiments can shape the space of possible mechanisms.

Consider one of the most famous experiments in the history of biology. Arthur Pardee, Francois Jacob, and Jacque Monod performed this famous series of genetic experiments in the 1950s. Biologists now refer to it as the PaJaMa experiment. In the 1950s the authors were investigating a puzzling phenomenon of enzyme induction. When other food sources are present and no milk is present, E. coli bacteria do not synthesize the enzymes needed to digest lactose, the sugar in milk. However, if bacteria are given milk as the only available food source, then the bacteria begin to synthesize these enzymes. Lactose is therefore said to be the inducer for these enzymes. One of the three induced enzymes is β-galactosidase (β-gal for short). The gene that produces it is called, appropriately, β-gal. The normal β-gal gene is designated by the symbol z^+; a mutant form that doesn't produce a functional enzyme is z^-. The researchers hypothesized a type of mechanism schema for induction: the lactose serves as an inducer that turns on (by some unknown activity) the inducible genes.

To their surprise the researchers found a mutant strain of bacteria that produces the enzymes constitutively, that is, even when milk is not present. Nature had intervened to produce a variant of the phenomenon. This is just what bacterial geneticists need: a normal and a mutant available for crossbreeding. The researchers hypothesized that this constitutive mutant has a previously undetected mutant gene, i -, that produces an endogenous inducer that functions just as lactose does in the inducible E. coli. That is, it produces its own inducer that triggers the expression of the β-gal gene.

To test the induction schema they engineered a mutant bacterial strain and conducted genetic experiments. They took an i- mutant (which generates continuous synthesis of the β-gal enzyme), but they replaced the normal z^+ gene for this enzyme with the z^- mutant. According to the induction hypothesis, the resulting bacterium would not be able to produce β-gal even though the endogenous inducer is present in the cytoplasm. They also isolated a normal $i^+ z^+$ strain of bacteria, with functional inducible and β-gal genes, that is, as usual, inducible in the presence of lactose.

Using these prepared strains they performed two reciprocal bacterial mating experiments. When E. coli bacteria mate (technically, conjugate), a small bridge forms between a male (donor) cell and a female (recipient) cell. The male's DNA

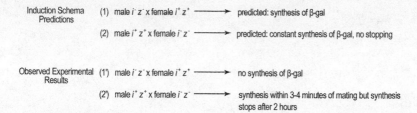

Induction Schema Predictions

(1) male $i^- z^-$ x female $i^+ z^+$ ⟶ predicted: synthesis of β-gal

(2) male $i^+ z^+$ x female $i^- z^-$ ⟶ predicted: constant synthesis of β-gal, no stopping

Observed Experimental Results

(1') male $i^- z^-$ x female $i^+ z^+$ ⟶ no synthesis of β-gal

(2') male $i^+ z^+$ x female $i^- z^-$ ⟶ synthesis within 3-4 minutes of mating but synthesis stops after 2 hours

Figure 8.5 Predicted and observed experimental results of the PaJaMa experiment.

(a single circle in E. *coli*) is first copied and then transferred linearly across the bridge into the female cell's cytoplasm. Figure 8.5 shows the predicted and actual results of these crosses.

The induction schema predicted that in experiment (1), when the i^- gene enters the female's cytoplasm, it forms the presumed endogenous inducer, which induces the normal z^+ β-*gal* gene to begin synthesizing the enzyme. But the observed results (1') did not match this prediction: the bacteria did not synthesize the enzyme. For the reciprocal cross in experiment (2), they predicted on the basis of their hypothesized induction mechanism that when the functional z^+ β-*gal* gene entered the female's cytoplasm, where the i^- gene was producing endogenous inducer, synthesis would begin and continue without stopping. The observed results (2') were that the bacteria began synthesis but that it stopped after about two hours.

The induction schema was thus disproved, and they had to construct a new mechanism schema to account for the observed results. Monod recounts their reasoning at this key juncture:

> Of course I had learned, like any schoolboy, that two negatives are equivalent to a positive statement, and [we] debated this logical possibility that we called the "theory of double bluff," recalling the subtle analysis of poker by Edgar Allan Poe. . . . I had always hoped that the regulation of "constitutive" [continuous synthesis] and inducible systems would be explained one day by a similar mechanism. Why not suppose, then, . . . that induction could be effected by an anti-repressor . . . ? (Monod 1965, p. 199)

They thus proposed that the inducer acts not directly, as in the induction schema, but rather by inhibiting an inhibitor (or via what is sometimes called double prevention). This was a discovery of a new kind of mechanism for gene regulation, a derepression mechanism. The i^+ gene produces a repressor, which prevents synthesis of β-gal by binding to the DNA immediately upstream from z^+. When lactose enters the cell, it binds to the repressor, which then falls off

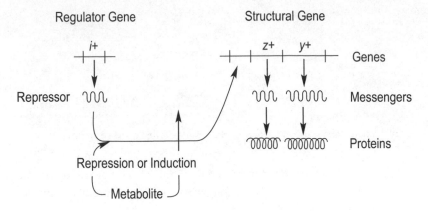

Figure 8.6 Lac operon schema of repression and derepression (induction).

the DNA and allows synthesis to begin. The mutant i⁻ strain does not produce a functional repressor, so synthesis is not prevented and occurs constitutively. The inducible and mutant cases are each explained by the same mechanism of repression/derepression, as Monod wished. See the schema in Figure 8.6.

We see how this mechanism schema accounts for the observed experimental results. In experiment (1) the female has a functional i⁺ gene that produces the repressor, so no synthesis occurs. This is the normal repressed state, and the introduction of the male's genes has no effect. In experiment (2), when the male's functional z⁺ gene enters the i⁻ z⁻ female, in which no repressor is present, that gene begins synthesizing the enzyme. (That it began so quickly was a puzzling anomaly, as we'll discuss in the next chapter.) Subsequently, the i⁺ gene enters the female and synthesis of the repressor begins. Within two hours it builds up in sufficient concentrations to block synthesis of the milk-metabolizing enzymes.

This experiment, one of the most important experiments in the history of genetics, ushered in an entirely new way of thinking about the mechanisms by which organisms regulate gene expression. Yet it would distort the structure of the experiment to see it as an instance of an experiment for testing causal relevance (where we intervene on one variable and detect another) or as an experiment driven by the goals of identifying components in a mechanism (which have roughly the same structure). Rather, the significance of the experiment lies in its ability to test a hypothesis about the active organization of the mechanism, to reveal that the mechanism involves the inhibition of an inhibitor.

Like Axelrod's work on the deactivation of norepinephrine, the PaJaMa experiment involves multiple interventions. Interventions are used to produce the required mutants, to put them in the same environment, and to entice them to conjugate with one another. The slow entry of the DNA of one bacterium into the other allowed sequential interventions, one for the i^- genes, and one for the z^- genes. And the timing of these interventions was absolutely crucial if the results were to be used to reveal something about the active organization of the mechanism. In experiment (2) the immediate production of β-gal eventually stops once the i^+ gene has entered the female and once enough time has passed for concentrations of the repressor to become high enough to shut down β-gal production. The experiment is thus not a simple intervene-and-detect kind of causal experiment, but a complex of interventions, organized and timed in such a way that they reveal something about the organization of the mechanism.

It is also significant that this experiment fundamentally involves engineering the experimental systems in which the experiment takes place. The system includes not merely a sample of the wild type bacteria but a contrived double mutant that allowed the precise interventions described above to distinguish the roles of the gene that produces the repressor from those for the milk-utilizing enzymes. In contemporary biology, many experiments take place in experimental systems that have been bred and cloned for properties that make them especially useful for experimental purposes. In many cases, this engineering is not antecedent to the experiment, something readily cordoned off as a background against which the primary intervention takes place. Instead, the preparation of experimental systems is itself crucial to understanding just how the experiment works. It is only in the context of such contrived experimental systems that the proposed interventions can meaningfully be interpreted. Model organisms and model systems are rightly prized by biologists precisely because standard organisms and systems have been prepared in such a way as to afford the researcher a particular kind of leverage over the mechanism that could not be achieved or presumed without the active construction of the model system.

CONCLUSION

The ascendancy of the mechanical worldview in the sixteenth and seventeenth century was paralleled by a vision of science, championed by Francis Bacon, according to which nature has to be vexed and hounded into to giving up her secrets. The metaphor is no longer so appealing. Nonetheless, the idea that we learn from nature by manipulating it and detecting the consequences of those interventions remains fundamentally correct. Perhaps it would be better to think of experimentation as an interview with nature. We pose questions in the form of interventions and contrived experimental arrangements, and we get answers through our detection apparatus. It is left to the scientist to ask just the right questions and to decipher what the world has to communicate.

In this chapter we discuss several kinds of experiments. The first two kinds, experiments for testing causal relevance and experiments for bridging levels of mechanisms, are regularly taught to students in introductory science classes. The last set of experiments for testing mechanisms tells us something more interesting about the nature of the experimental interview. In particular it tells us that different kinds of questions about a mechanism are answered with different kinds of experimental strategies. It tells us that the conversation is often protracted, involving multiple interventions and series of experiments. It also tells us that the subject of this interview, the experimental system, often must be prepared before it can deliver answers to our most insightful questions.

In seeking to understand methods for discovering mechanisms, we begin to think more inclusively about the many kinds of experiments one finds across the life sciences. Some experiments are designed to learn what the activities of a mechanism are; others address the entities. Some experiments make sense only in the context of a series of experiments, compiled to reveal something about the organization of a mechanism. Some experiments involve engineering experimental systems to produce mechanisms with specific properties that allow us to use them to reveal surprising facts about how a target mechanism works.

In the *Novum Organum*, Bacon describes a variety of experimental methods by which scientists might discover causes and produce effects, as he defined the goal of natural philosophy. If one takes the search for mechanisms seriously, one might continue Bacon's project by reflecting on the ingenious experimental designs by which scientists convince nature to reveal its mechanistic secrets.

BIBLIOGRAPHIC DISCUSSION

Francis Bacon propounded his vision of a new method for (what we now call) science in *The New Organon* (1620). The intervene-detect form of experimentation on mechanisms is discussed in more detail in Woodward (2003) and Eberhardt et al. (2010). Multilevel experiments in neuroscience are discussed in Craver and Darden (2001), Craver (2002; 2005; 2007, ch. 4). For more on Goldberger, see Bollet (1992). For more on Otto Loewi, H. H. Dale, and the context of their research on neurotransmitters, see Valenstein (2005). Julius Axelrod's work is discussed in Craver (2008b) and originally appeared in a series of papers by Axelrod and his collaborators in 1961. Bender et al. (1984) discuss the methodology of perturbation experiments in community ecology. Forterre et al. (2005) provide evidence for the way the Venus flytrap mechanism operates.

Oswald Avery's paper on bacterial transformation, written with C. M. MacLeod and Maclyn McCarty, was published in 1944 (Avery et al. 1944). The definitive biography for Avery is Dubos (1976). McCarty (1985) gives an autobiographical account of their discovery.

The PaJaMa experiments that resulted in the discovery of regulatory genes and messenger RNA are detailed in Pardee et al. (1959) and recounted in Pardee (1979), who calls it PaJaMa. Schaffner (1974; 1993) discusses the example in the context of thinking about discovery and reduction in the biomedical sciences. Schaffner (1974, footnote 42) notes that the PaJaMo or often PaJaMa takes its name from the first letters of the authors' names: Pardee, Jacob, and Monod. He notes that the pajama name is thought to be especially appropriate because the experiment is a bacterial *mating* experiment. However, Schaffner continues, who first used the amusing name seems not to be known. The case is also discussed in Darden and Craver (2002).

9 STRATEGIES FOR REVISING MECHANISM SCHEMAS

INTRODUCTION

The engine of mechanism discovery is in many cases fueled not by experimental successes, in which a prized mechanism schema is confirmed by careful observation and experiment, but by failures in which the prized schema turns out to contain an error. So while there is a tendency to despair at moments when the world does not cooperate with our favored schemas, such failures are also a cause for celebration: failures often contain within them clues that guide the construction of more plausible schemas.

Mechanism schemas can fail in many ways. Understanding just how they fail is often a first step in building better schemas. If the schemas fail because they are incomplete, for example, the black and gray boxes in a sketch might guide research to fill them (as discussed in Chapter 6). If they fail because they start from an incorrect characterization of the phenomenon, then one might, for example, split what was once considered to be one phenomenon into two phenomena or, alternatively, one could recharacterize the phenomenon as an instance of a more inclusive type (as discussed in Chapter 4).

In this chapter, we focus on problems that arise when one finds an *anomaly* for one's mechanism schema. An anomaly is an empirical finding that appears to conflict with a how-possibly/how-plausibly schema. The empirical finding may arise from a specific test of a prediction of the schema or from exploratory experimentation and observation. However the finding is unearthed, such a finding becomes an anomaly when someone recognizes that the result violates an expectation, or prediction, that follows from a schema. In such cases, the scientist knows that something has to be changed but faces a choice among a number of possible responses. One might choose to question the empirical finding, its replicability, or the methods used to produce it. One might, however, see it as a problematic finding for one or more parts of a hypothesized schema. Anomaly resolution is the process of figuring out how to correct the mismatch between a schema and one's empirical findings.

In resolving an anomaly one must diagnose the type and location of the failure, and for some types of anomalies propose a redesigned module at that fault site in the schema to replace the failing one. Physicians have strategies for diagnosing diseases and targeting remedies at appropriate physiological processes. Computer scientists have codified strategies for writing, testing, and debugging software. Engineers study principles of design and learn how to redesign artifacts that don't work or that break down. Analogous diagnostic, testing, and redesign strategies are useful for resolving anomalies for proposed mechanism schemas.

In this chapter, we develop a taxonomy of types of anomalies and the strategies for resolving them. Some types of anomalies indicate a problem with the experimental results; others suggest problems with the hypothesized schema. Below is a list of the key types of anomalies, each of which is then discussed in the following sections.

GENERAL TYPES OF ANOMALIES

- **Experimental Error**
- **Data Analysis Error**
- **Monster Anomaly**
- **Special Case Anomaly**
- **Model Anomaly**
- **Falsifying Anomaly**

These categories of anomaly are arranged in order from less serious to more serious problems for the how-possibly/how-plausibly schema under test. The list begins with two types of anomalies that require no change to the schema: namely, experimental and data analysis errors. It ends with the kind of anomaly that requires abandoning the schema and searching for another (or another type of) schema, namely a falsifying anomaly. In between the categories of experimental error and falsifying anomalies are kinds of anomalies that require diagnosis and redesign. As we will show, the mechanistic framework discussed in previous chapters provides guidance in how to resolve monster, special case, and model anomalies.

EXPERIMENTAL AND DATA ANALYSIS ERROR
AND FALSIFYING ANOMALIES

Methods for credentialing experimental results and doing data analysis are commonly taught to science students in methods and statistics courses. We mentioned some of these in earlier chapters: find tractable experimental systems in which the target mechanism is running, set up proper controls, intervene to change a single mechanism component, detect the result of the intervention, and reproduce the experiment to be sure the results are accurate. Some kinds of tests require randomization. Statistical tests require attention to adequate numbers of experimental subjects and trials, as well as principled ways of dealing with outliers (incongruous data points). These methods and safeguards for doing science in general also apply to testing predictions of mechanism schemas. We simply endorse them here. If the anomaly is due to experimental error, then no change in the proposed mechanism schema is needed, and the anomaly is resolved. On the other hand, if the experimental result is shown to be an accurate report of something about the mechanism running in the experimental system, and it appears to be in conflict with the proposed schema, then further work is required to resolve the anomaly.

At the other end of the spectrum of severity for the proposed schema are falsifying anomalies. Sometimes an experiment is designed to choose among two or more how-possibly/how-plausibly schemas. A well-designed and well-executed *crucial experiment* yields a falsifying anomaly for one (or one type of) how-possibly schema, while providing positive evidence for the rival. The anomalous result rules out a portion of the space of possibilities and makes the rivals more plausible. Systematic consideration of alternative schemas and empirical work to sort them from one another is an important part of schema evaluation.

In other cases, where only one schema is under test and it is already plausible because of prior supporting evidence, many anomalies may be required before the schema is abandoned. A single anomaly for an otherwise plausible schema may be put aside, pending further developments. Why, after all, should a single experiment kill a beautiful theory? Disagreements about the severity of the challenge are a healthy part of the way the community of scientists operates. Advocates of the schema may try to resolve the anomaly by diagnosing it as belonging in other less severe categories, while critics may take the opposite line. Nonetheless, when scientists reach agreement that an anomaly or set of anomalies falsifies a schema, then a portion of the space of possibilities has been ruled out. If no rivals are available, then one has to construct and search another portion of the space of possibilities by, for example, finding another type of schema, putting together new modules, or forward and backward chaining (as discussed in Chapter 5).

MONSTER, SPECIAL CASE, AND MODEL ANOMALIES

Intermediate in severity between experimental error and falsifying anomalies are types of anomalies that seem to require a change within a hypothesized schema. *Monster anomalies* at first seem to impugn the schema but, in the end, can be blamed on factors outside of the mechanism of interest. In our example below, an anomaly seemed to indicate that something is wrong with one or more of the basic schemas for explaining heredity. In fact, however, the anomalous finding resulted not from a problem with the genetic schemas but with a fault in a later, embryological, stage of development. *Special case anomalies* also appear to impugn the schema but, in the end, can be blamed on the fact that one's experimental system is a special case exception. The schema is not wrong; rather, it has exceptions, and the experimenter has identified one of those exceptions. For example, the finding that some viruses use RNA as their genetic material, while other viruses and all living things use DNA, does not show that the central dogma is false. Rather it shows that some organisms use RNA instead of DNA as their hereditary material, that is, that the central dogma has a more restricted domain than it was previously thought to have. A *model anomaly*, however, is a more serious concern. In this case, the schema fails because it describes the target mechanism incorrectly. The fault cannot be passed on to some other mechanism (as it is in monster anomalies) and it cannot be attributed to a special-case exception. Rather, the anomaly suggests that one must begin to revise the schema for the entire domain of the model. In the history of genetics, for example, the early Mendelians' assumption of independent assortment was found to have numerous failures. Some groups of characters are linked in inheritance, while others assort independently. Resolution of the anomaly showed that genes along the same chromosome are linked in inheritance. This is the normal or model case: for all species with chromosomes, genes occur in linkage groups.

When an empirically credentialed and non-falsifying anomaly falls into one of the three categories of monster, special case, and model anomalies, then the mechanistic framework provides guidance for anomaly resolution. It helps to categorize the anomaly. It helps to localize the failure outside (for monster anomalies) or within the hypothesized schema (for special case and model anomalies). It helps researchers to redesign a failing schema. It helps researchers to design experiments to find the scope of a newly redesigned schema. Diagnosing the location of failures and redesign of mechanism schemas are the subject of following sections.

CLUES FROM THE ANOMALY
A first step in anomaly resolution is to characterize the nature of the anomaly by comparing the anomalous result with what was predicted or expected given the hypothesized schema (assuming that the anomalous result has been credentialed through, for example, replication). The goal of such an analysis is to mine the anomaly for clues about why the hypothesized schema failed.

Assume that the schema predicts P and that empirical investigation reveals that O, rather than P, is the case. The first task is to see how P and O differ from one another. The nature of the discrepancy may provide guidance during anomaly resolution. One might expect the mechanism to produce something that is not in fact produced (as the example below, in which the expected AA double dominant genotype is missing, illustrates). This is a missing element anomaly. One way to resolve such an anomaly is to trace the path that should have produced the missing element to localize one or more possible sites of failure. We discuss other types of anomalies below.

A MONSTER ANOMALY

Localizing the anomaly is a matter of figuring out what in a schema has gone wrong. For simplicity, assume that one is investigating a linear mechanism (that is, one that works sequentially, without cycles, from beginning to end), and that this mechanism can be situated in a wider temporal context, with other clearly individuated mechanisms operating before and after it. Such a situation is represented in Figure 9.1. (We neglect for present purposes the discovery of cyclically organized mechanisms, only because to our knowledge, *general* strategies for dealing with cyclical mechanisms have yet to be developed. The search for such strategies is an active research area in machine learning, systems biology, and philosophy).

A monster anomaly locates the failure outside the target schema A→E, in this case, in Mechanism 0 or in Mechanism 2. The name for this category—monster anomaly—is meant to suggest that the anomaly reflects a deviant case. A monster anomaly appears at first to impugn the schema, but it does not. The schema need not be revised because the fault is localized outside the target mechanism. The experimental result is barred as a monster.

An example of a monster anomaly comes from the early days of Mendelian genetics. In a cross between heterozygous yellow mice (Aa x Aa), the expected ratio of double dominants to hybrids to double recessives is 1:2:1 (AA:2Aa:aa). Lucien Cuénot (1866–1951) reported in 1905 that his efforts to breed heterozygous yellow mice failed to deliver the predicted ratios. Instead, he found a ratio of 2:1 (2Aa:aa). He found no pure yellow mice. The 2:1 ratio is a missing element anomaly; some characteristic aspect of the phenomenon is absent. The anomaly

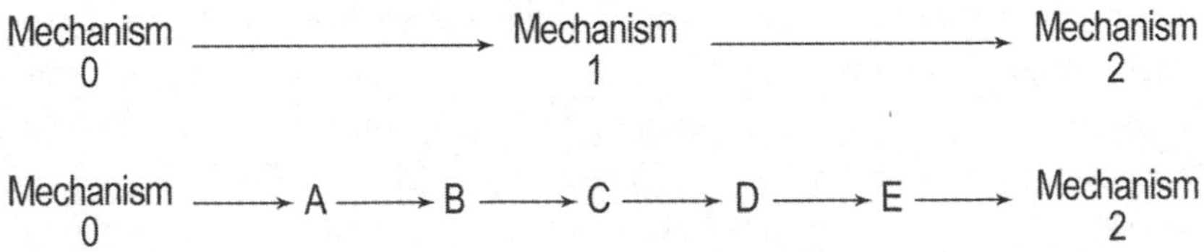

Figure 9.1 Mechanisms before and after a target schema.

Mendelian Segregation

Consider a breeding experiment between two hybrid heterozygote forms:
Aa x Aa.
A is the dominant form and produces yellow mice; *a* is the recessive form.

The alleles segregate so that each germ cell has one or the other, but not both, of the two forms of allele.
This is often called "purity of the gametes" and is sometimes considered to be the essence of Mendelian genetics.
The alleles are thus either *A* or *a*; each sperm and egg has one or the other but not both.

The next stage is the random mating between the two forms, often represented by a Punnett Square:

gametes	A	a
A	AA	Aa
a	Aa	aa

This results in 1*AA*:2*Aa*:1*aa* ratios of the genotypes.
AA is the double dominant; *Aa* the heterozygote; *aa* the double recessive.

If the phenotype of the pure dominant (AA) and heterozygote (Aa) are alike, then 3:1 is the phenotypic ratio.

Table 9.1

is the discrepancy between the predicted (P) 1:2:1 ratio and the observed (O) 2:1 ratio. The double dominants, *AA*, were missing.

Another lab group replicated these results; the anomaly was likely not the product of experimental error. It appeared that the schema for Mendelian segregation had to be revised. The normal stages of the mechanism of Mendelian segregation are depicted in Table 9.1. Each stage of the mechanism schema presents a potential opportunity for failure.

To resolve the 2:1 anomaly, different geneticists proposed different locations within this schema and proposed different redesign hypotheses. One can consider each stage as a possible candidate for localization and redesign. Lucien Cuénot localized the problem to a stage after the formation of gametes, in the fertilization stage. Perhaps a selective process that occurs during fertilization precludes random mating between all types of gametes (called selective fertilization). As the Punnett square in Table 9.1 shows, under normal circumstances one expects the two possible kinds of sperm (A and a) to combine randomly with two different kinds of eggs, yielding the 1:2:1 ratio. Perhaps in this case, Cuénot proposed, something prevents the AA combination from occurring. If so, this would be a special case anomaly: random fertilization usually occurs, but selective fertilization happens in special cases, such as in this strain of yellow mice (and some other alleged cases that Cuénot mentioned).

Thomas Hunt Morgan (1866–1945) proposed a more drastic response to the 2:1 anomaly. Morgan later became an advocate of Mendelism, but in 1905 he was a skeptic about pure Mendelian segregation. Indeed, it is puzzling that pure dominant and recessive forms can emerge from a hybrid cross and never again show the effects of their hybrid parents. It was an old idea that mongrels show the effects of their ancestry. To account for this anomaly, Morgan proposed that pure segregation never occurs. His hypothesis would have required that gametes be represented not by A or a alone but by A(a) and a(A), showing that pure segregation never occurs. Morgan suggested that Cuénot had found no pure breeding dominants because the mongrel effect appeared earlier than usual in this case. If Morgan's view had been correct, then the 2:1 anomaly would have been a falsifying anomaly, showing that the purported segregation of pure alleles (shown in Table 9.1) never occurs. If this view had prevailed, and if purity of the gametes is taken to be the essence of Mendelism, then Mendelian genetics would have been disconfirmed.

Finally, William Castle (1867–1962) and his colleagues, who replicated the mice study to rule out experimental error, localized the problem to yet a later stage of the mechanism. According to Castle, the A and a type gametes form, random fertilization occurs, but the double dominant (AA) yellow embryos die. If so, then one would expect to discover undeveloped embryos for yellow AA mice in the mother's womb. Castle did dissections and found precisely that: dead yellow mouse embryos. In doing so he established that the source of the anomaly is downstream from the mechanism of Mendelian segregation and located among the details of embryonic development.

No methods at the time could isolate types of germ cells in mice to see if they contained pure alleles. Nor were there techniques for analyzing the fertilized eggs to see if the fertilization was, in fact, random. Those stages were not experimentally accessible. But Castle had all the technique and skill he needed to dissect out the embryos and confirm their yellow color. Castle's test involved chaining backwards through the stages of a mechanism to find a stage at which available techniques might plausibly shed light.

Castle's experiment confirmed his hypothesis and resolved the anomaly, removing the problem for Mendelian genetics. The 2:1 ratio was a monster anomaly, localized downstream from the mechanisms of Mendelian inheritance.

The AA combination was lethal at some stage of embryological development. A new category of failures was added to the geneticists' repertoire for explaining why 1:2:1 ratios are sometimes not to be found. From that point on, it was part of the compiled knowledge of the field that a failure to achieve the expected Mendelian ratios might be diagnosed as a lethal gene combination rather than some altogether different hereditary mechanism. Embryologists had the task of finding out why the normal embryological developmental mechanisms fail and why that particular gene combination kills the mice embryos. But this task was not especially pressing for geneticists, who could cordon off the problem as irrelevant to the target mechanism they hoped to understand. Monster anomalies like this fail to threaten the schema for a mechanism because the anomaly can be explained by something that happens in an altogether separate and distinct mechanism. Castle's work saved the Mendelian schema by barring this missing element anomaly from the domain of the schema for Mendelian inheritance. Thinking of a mechanism within its wider temporal context, where other mechanisms operate before and after it, can help one to find candidate locations for the monster anomaly.

LOCALIZING AND FIXING SPECIAL CASE AND MODEL ANOMALIES
In contrast to monster anomalies, special case and model anomalies are localized within the schema. Such anomalies force one to identify which parts of the schema are likely responsible for the discrepancy between the predicted and the observed result. Once the schema has been redesigned, it is necessary to determine the scope of the new schema. For special case anomalies, the domain is split; the new schema applies only to a small subset of items. For model anomalies, the newly redesigned schema applies to the entire domain.

Again consider the representation of a simple, linear, non-branching schema (as strategies for more complex cases are quite complicated and in some cases unknown):

$$A \rightarrow B \rightarrow C \rightarrow D \rightarrow E$$

Suppose a scientist finds O (rather than E) at the end of this mechanism. Suppose further, the scientist has ruled out experimental error and suspects something has gone wrong within the schema itself. That is, suppose monster barring is not an option. The scientist now has to determine what has gone wrong within the schema. Assuming a linear mechanism, one efficient search procedure is to start in the middle and to see, via further empirical investigation, whether the schema is correct about C. If the schema is correct about C, then (given our assumptions about this mechanism), the fault must lie between C and E, namely in or around stage D. If, on the other hand, the scientist determines that C is also incorrect, then they are directed backwards to see if B occurs.

Starting in the middle (binary search) is an efficient search procedure for localizing an anomaly in a linear mechanism, if all the stages of the mechanism are equally implicated as possible sites of failure, and if all the states are equally accessible to investigation. However, this may not be the case. The nature of the discrepancy between predicted and observed outcomes might implicate some sites and not others. Investigative methods may provide access to some stages but not others. So, other search procedures suggest either starting at the end and working back (as we saw with Castle's work on yellow mice), or starting at the beginning and working forward. The goal is to find out where the schema succeeds, where it fails, and where the fundamental source of the error lies. The assumption that the mechanism exhibits modular (that is, separable) and linear (that is, sequential) stages aids that search by allowing one to work forward or backward to localize the failure site. (Of course, more complex search strategies are needed for mechanisms that do not exhibit this kind of organization.)

As one begins to investigate potential sites of failure within one's mechanism schema, a number of other strategies are useful for determining just how the schema needs to be revised. Again, the nature of the failure offers signposts to the path of success. Furthermore, that the search is for a mechanism shapes the kinds of possible anomalies to be considered. Here are a few of the reasoning strategies by which a kind of anomaly indicates something about the sort of re-design that is likely to work. (Compare the evidential constraints during schema evaluation in Table 7.1.)

INDICATIVE ANOMALIES IN THE SEARCH FOR MECHANISMS

- **Entity Anomaly and Entity Change**

    If there is a problem with the entity, then either change its hypothesized structure, orientation, size, or activity-enabling property, or find a different entity or kind of entity.

- **Activity Anomaly and Activity Change**

    If that entity or module doesn't have the needed ability, then find a different activity or kind of activity. For example, if the assumed mechanism was hypothesized to have only excitatory activities, add inhibitory activities as well.

- **Temporal Anomaly**

    If there is a problem with timing, such as the order, rate, duration, or frequency, then consider what might speed up or slow down the action or whether different entities and/or activities are needed to work at this rate or frequency.

- **Role Anomaly**

  If (i) the gray box appropriately characterizes the function in the mechanism but the entity filling that role in the schema cannot or does not play that role, then find another entity or activity to play that role. Further specification of the functional role in question provides guidance in this search. If (ii) the gray box is wrong, then change the hypothesized function, and see if the hypothesized entities can fulfill that role or search for another occupant of that role.

- **Chaining Anomaly**

  If one stage does not have the expected productive relation to another stage (i.e., the output of a previous stage does not match the input to the next or vice versa, or the activity signature of a later stage is not as expected, given its relation to the previous stage), then change one or the other stage to bring them into an appropriate chaining relation to reveal the mechanism's productive continuity.

- **Organizational Anomaly**

  If there is a problem with the overall organization of the schema, then change the hypothesized organization. For example, for a missing stage, add a stage before or after the faulty stage. For a more serious organizational anomaly, then consider complicating an over-simplification. For example, if the mechanism was assumed to be linear, instead consider a fork and join, or a cyclic feedback, or a maintenance form of organization.

- **Integrative Anomaly**

  If the anomaly presents a problem that cannot be solved with the techniques and concepts of the field or at the level at which the schema was originally constructed, then consider going to another field or to another level to resolve the problem.

There is as yet no systematic treatment of how these different types of anomalies for mechanism schemas indicate different kinds of failures. Typically, there will be a number of candidates for correcting any given failure in a mechanism schema. As such, the above list should be understood as a set of redesign strategies that scientists have used or could use to resolve anomalies. Whether a given strategy works in a particular case, however, depends on the empirical details of the mechanism in question. Nonetheless, it seems to us a useful enterprise to begin sketching out the kinds of problem-solution pairs that have proved especially useful to scientists in key discovery episodes.

As just one example consider an episode in the discovery of the mechanism of protein synthesis. In this case we are focusing on a schema for which anomalies arose for the part of the schema in which the ribosome acts as a template for the ordering of amino acids in a protein. This single example illustrates many of the above problem-solution pairs: entity anomalies, role anomalies, temporal anomalies, and chaining anomalies.

In the mid-1950s molecular biologists hypothesized that the ribosome was the template for transferring the order of the bases in the DNA to the order of the amino acids in a protein. In this sketch the gray box of "RNA serving as a template," was filled with an entity, the ribosome. The schema is represented in Figure 9.2.

Cell biologists had shown that in eukaryotes (creatures with a separate nucleus and cytoplasm), protein synthesis is associated with the ribosomal particles in the cytoplasm. In the mid-1950s molecular biologists proposed a schema in which sequences of DNA bases function as a site for complementarily copying of template RNA. Then the template RNA moves away from the DNA, combines with other components, and forms a stable ribosome. They hypothesized that this template somehow determines the order in which amino acids are incorporated in a growing protein.

Molecular biologists knew a bit about the internal structure of these ribosomal particles. They knew that with proper staining these particles are microscopically visible and consist of two types of RNA, one larger and one smaller (as well as some other structural components). They assumed that each ribosome is specific to the particular protein for which it serves as a template. One hypothesis was that RNA forms a double helix (by analogy with the DNA double helix) and that differently shaped holes in the grooves of the helix serve as the place where different amino acids could fit (like a lock fits a key). The RNA lines up the amino acids, which then bond to form the protein.

Anomalies quickly began to accumulate for the ribosome-as-template module of this mechanism sketch. With a bit of hindsight, one can see the molecular

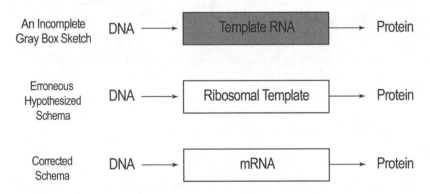

Figure 9.2 Sketch and schemas for the protein synthesis mechanism.

biologists using these anomalies to guide their revision of the protein synthesis schema.

Consider anomalies that arose as scientists attempted to grow proteins in a cell-free system. Biochemists combined DNA and radioactively labeled amino acids in a medium and, by adding other required ingredients, induced them to form into growing protein chains. This artificial system allowed biochemists to intervene on different stages of the mechanism and thereby test the roles of different components in the mechanism. One prediction of the schema is that the continued presence of DNA is not required for the formation of proteins. Once the DNA has made the ribosomal template, the DNA no longer serves any useful role. One could destroy the DNA at this point with no effect on protein synthesis. However, that prediction was not confirmed. Adding an agent to the system that destroys DNA while leaving ribosomes intact halts protein synthesis. This is a role anomaly and hints at an entity anomaly. It is a role anomaly because it suggests a function for DNA not represented in the schema. Once the RNA is transcribed from the DNA and the stable ribosome particle has been formed, the continuing role played by DNA is puzzling. This puzzle hints at an entity anomaly. Maybe there was some sort of previously unrecognized unstable intermediate between DNA and the protein that was being continuously transcribed from the DNA. Perhaps the properties of being long-lived and stable were not the ones needed for the entity to play the template role.

Another prediction of this sketch is that the proportions of bases in DNA and its complementary RNA should correspond to one another. If the DNA base sequence is complementarily copied (transcribed) to RNA, then one would expect that the ratios between the bases in DNA and the bases in RNA would be similar in a given species. In the late 1950s, experiments showed that no such correspondence existed. At first, the scientists were convinced that the anomaly must be due to an experimental error or a mistake in data analysis. However, this possibility vanished with repeated replication of the results. This chaining anomaly indicated a possible problem with the ribosome-as-template schema. DNA base ratios differ widely from species to species, but the RNA base ratios do not. This is a chaining anomaly because the expected relations between the stages were not found; the hypothesized output of one stage does not match what one learns about subsequent stages. Again, the anomaly was localized to the intermediate stage of the ribosome-as-template. Multiple anomalies seemed to indicate that the large stable ribosomal particle was not the appropriate component to fulfill its assigned role of template in the mechanism.

Other anomalies arose through the use of an in vivo system for manipulating protein synthesis in E. coli. E. coli bacteria sometimes mate (technically, they conjugate). A small bridge forms between a male (donor) cell and a female (recipient) cell. The male's chromosomal DNA is first copied and then transferred across the bridge into the female cell's cytoplasm. In the 1950s Pardee, Jacob, and Monod performed the PaJaMa experiment (discussed in Chapter 8) that revealed a temporal anomaly in the ribosome-as-template schema. They engineered an experimental system in which a gene was dysfunctional in the female but functional in the male. Specifically, the female could not synthesize an enzyme used to digest the sugar in milk (lactose). In contrast the male had a functional gene for that enzyme. Given the ribosome-as-template hypothesis, the prediction was that when the two mated, protein synthesis would not begin quickly. After the gene from the male entered the female's cytoplasm, it would take some time for that gene to be copied to ribosomal RNA, for the large stable ribosomal particle to form, and for the required enzymatic protein to be synthesized. That prediction was not confirmed. Rather, the female bacterium synthesized the enzyme within 2 to 3 minutes after the functional gene in the DNA from the male entered the female cytoplasm (as we showed in Figure 8.5). This discrepancy is a temporal anomaly. This anomaly also localized to the ribosome-as-template site in the mechanism sketch.

These mounting anomalies for the protein synthesis mechanism sketch all localized to the ribosome-as-template site. They also provided constraints that aided in redesign. The temporal anomaly indicated that the template should be synthesized more rapidly than the large ribosomal particles were known to be synthesized. The role anomaly indicated that the DNA should have an ongoing role in the mechanism even after the template has been produced. The chaining anomaly indicated that there should be a complementary relationship between the bases in DNA and those in the role player for template RNA. These constraints come from analyzing the anomalies for clues. The nature of these discrepancies between what was expected (P) and what was found (O) gave the scientists hints about the properties of the part needed at the site of the template. The ribosome-as-template component of the how-possibly mechanism sketch needed to be revised. The question was: what should play the functional role of template if the ribosomal particle did not?

MODULAR REDESIGN

After localizing the anomaly and analyzing the anomalies for constraints, the next task is to redesign the failed module. Engineers diagnose the site of failure in artifacts and then redesign the part that has been regularly failing. Redesign in engineering is a more constrained generation process than designing an entire artifact from scratch. The redesigned modular part has to avoid the kind of failure evident in the old design. Furthermore, the newly redesigned part has to work with the other parts of the device that haven't failed and don't need to be redesigned.

Analogously, when a model anomaly has been localized to indicate failure of one or more parts of a how-possibly schema, then that hypothesized module has to be redesigned. The strategies for constructing a mechanism schema (discussed in Chapter 5) are useful for doing redesign, but they are now more constrained. The nature of the anomaly and the nature of the previous failure together rule out part of the space of possible modules—those too much like the old faulty one in the relevant respects. Also, when an anomaly is localized to one part of the schema, the construction process is further constrained: the redesigned module must still work with (process inputs to and generate outputs for) the other parts. In other words the module must be redesigned to fit into the stages before and after it to maintain the productive continuity of the mechanism.

To return to the protein synthesis example: scientists attempted to redesign the how-possibly sketch at every possible subcomponent. To account for the anomaly of rapid protein synthesis in the bacterial mating experiment, it was proposed that bacteria might be a special case. In prokaryotes the protein might be synthesized directly on the DNA as it enters the cytoplasm of the female. This hypothesis uses the strategy of deleting the problematic step for a subset of the domain. Had this redesign candidate been correct, then this would have been a special case anomaly. The anomaly arises only in bacteria. The correct schema for bacteria (as a special case) would then have been as follows:

DNA→protein

Molecular biologists, in contrast, retained a general role for RNA, and generated a variety of candidate redesign hypotheses to resolve this chaining anomaly. The anomalies were localized at each of the molecules within the faulty part: DNA→RNA. Perhaps, they reasoned, only part of the DNA carries genetic information for proteins. Or perhaps only some subset of the RNA in the ribosomal particles (known to contain two separable RNA molecules of different sizes) carries a complementary copy of the genetic information, while other bits of RNA play a structural role. Both of these redesign hypotheses use the reasoning strategy of delineate and specialize: what was thought to be a property of the whole is instead a property of only one of the parts. Thus, one of the mechanism's components (either DNA or RNA) was split in two, with the functional role of coding region or template assigned to only one of the parts of the DNA or one of the parts of the ribosome. If so then the ratios of bases in the entire

DNA and entire RNA fractions would not be expected to correlate so tightly with one another.

A final candidate redesign hypothesis for this anomaly was that there is an as yet undiscovered template RNA, which is smaller than the ribosome and synthesized rapidly on the DNA after it enters the cytoplasm of the female. Instead of solving the problems by splitting the domain and deleting a component (as the bacterial DNA hypothesis did) or tweaking properties of known components (as the one-component-of-ribosome hypothesis did), molecular biologists posited a new and undetected component to fill the gray box of template. This newly proposed, rapidly synthesized RNA satisfied the constraints from both the rate and the chaining anomalies. Separate experiments had already provided evidence of a DNA–like RNA in viral infections. The question arose: could that be the missing type of RNA for normal cases in prokaryotes and eukaryotes (viruses are neither)? Was this special case result in viruses a model system for the normal result in prokaryotes and eukaryotes?

Only additional testing could choose among the alternative redesign hypotheses. The molecular biologists devised experiments to detect the existence of the rapidly synthesized messenger RNA (as it came to be called). Experiments differentially labeled the older stable ribosomes and components for a possibly newly synthesized messenger RNA. Such activation tracer experiments confirmed the new rapid synthesis of a smaller, less stable RNA. Rapid synthesis of messenger RNA proved (with further testing) to be how the normal mechanism of protein synthesis works in some viruses and in both bacteria and eukaryotes. The anomalies for the ribosome-as-template hypothesis were thus model anomalies. Their resolution led to the discovery of a new entity involved in transcription, namely messenger RNA. The resolution of this model anomaly showed how that portion of the normal protein synthesis mechanism works in *general*.

To summarize: the nature of the anomaly provides clues as to where the site of failure is localized within a how-possibly mechanism schema or sketch. Forward and backward chaining are particularly useful for finding candidate failure sites. The nature of the anomaly (or features in a set of anomalies) constrains the search for a new component or module to replace the failed one. In special case anomalies the newly redesigned schema/sketch applies only to a subset of the original domain. In contrast, model anomalies show what is normal for the entire domain and result in the discovery of a more general schema.

CONCLUSION

The chapter discusses reasoning strategies to resolve anomalies. We discuss several kinds of anomalies: experimental error, data analysis error, monster, model, special case, and falsifying anomalies. Monster anomalies do not require one to change the proposed mechanism schema or sketch. In special case and model anomalies, the scientist must localize the failure in a proposed schema or sketch and redesign the schema as needed.

A special case anomaly requires splitting the domain into two parts, with two different schemas to characterize the mechanisms that operate in the two domains. A model anomaly is resolved by changing the schema for the entire domain. The more specialized types of mechanistic anomalies (entity, activity, temporal, chaining, and so on) also suggest strategies for localizing the fault within a mechanism schema as well as new ideas for redesigning the schema.

Confirming evidence, it is true, encourages researchers to continue pursuing their favored how-plausibly models. Disconfirming evidence, an anomaly, has within it the power to change how scientists think. This is true not just because anomalies force scientists to revise their schemas but, in addition, because the anomalies, conjoined with what else one knows about the mechanism, help one to localize sources of failure within the schema and to redesign the schema to accommodate the anomaly. After redesign, the new schema enters the intellectual world of science more fit than its parents and so perhaps somewhat more likely to survive and influence the content of future schemas.

BIBLIOGRAPHIC DISCUSSION

Darden (1991) discussed monster and model anomalies; Darden (1995) added special case anomalies to the list. Burian (1996) added falsifying anomalies, which he called Kuhnian anomalies, following Thomas Kuhn's (1962) discussion of crisis anomalies that lead to a large scale (paradigm) change. (We have no example of such a large scale change, based on a single anomaly.) For more on anomalies and error probing in experimental results by philosophers of science, see Mayo (1996); Allchin (1997; 1999; 2002); Elliott (2004; 2006). Imre Lakatos (1976) named and discussed the reasoning strategy of monster barring in his work on providing mathematical generalizations. A strategy to save a generalization from a purported exception is to bar the exception as too abnormal (the monster) to require changing the definition, such as a polygon with a hole in the middle. Curiously, when Lakatos (1970) discussed reasoning in science rather than mathematics he talked about research program-specific negative and positive heuristics; he did not attempt to formulate more general reasoning strategies for revising a theory in the face of a purported falsifying instance.

For additional work connecting scientific discovery and reasoning in diagnosis, see articles in Schaffner (1985). Researchers in artificial intelligence have also addressed reasoning strategies for resolving anomalies; see Karp (1989;

1990) and Bridewell (2004). For more on the analogy between, on the one hand, diagnosing faults in mechanism schemas and redesigning them, and, on the other, diagnostic and redesign reasoning in engineering, see Goel and Chandrasekaran (1989) and Petroski (1992).

Darden (1991, ch. 8) discussed Cuénot's yellow mice monster anomaly in detail; primary sources include Cuénot (1905), Morgan (1905), Castle (1906), and Castle and Little (1910). For primary sources for the ribosomal anomaly, see Crick (1959), Watson ([1962] 1977) and for the discovery of messenger RNA, see Brenner et al. (1961), Gros et al. (1961), and the autobiographical accounts in Crick (1988) and Jacob (1988). Secondary accounts include Judson (1996), Rheinberger (1997), Darden and Craver (2002), and Olby (2009).

10 INTERFIELD AND INTERLEVEL INTEGRATION

DIFFERENT FIELDS, DIFFERENT QUESTIONS

Biology comprises many fields, including anatomy, physiology, embryology, ecology, evolutionary biology, genetics, molecular biology, and neuroscience. These different fields of science are distinguished from one another by the fact that they typically address different central problems, use different investigative techniques, and describe phenomena with different vocabularies. Sometimes they address entirely distinct domains of investigation.

Often, however, different fields of biology address fundamentally different kinds of questions about the same phenomenon. Take the sense of touch in the tiny roundworm *Caenorhabditis elegans*. An evolutionary biologist might be interested in how the features of touch in the worm came to be shaped by selection in ancestral environments. A developmental biologist might be interested in how the sense of touch develops over the early days of the worm's life. A geneticist might be interested in learning which genes contribute to the development of a functioning haptic system. A neuroscientist might be interested in the neural circuitry that mediates the response to touch. A molecular biologist might be interested in the structure and movements of the ion channels that gate the touch signals. A comparative anatomist might be interested in the differences between the sense of touch in roundworms and the sense of touch in flatworms. Different phenomena call out for different mechanistic explanations, and so different fields are often engaged in different explanatory projects.

Often, however, biologists find it necessary to integrate what is known from the perspective of one field with what is known from the perspective of another. To understand a mechanism, one typically needs to know many different things about many different kinds of components. One needs to know about the entities involved, their various properties, the activities in which they engage and their spatial and temporal organization. Often different techniques and experimental practices are required to collect these constraints. An anatomist, for example, might be interested in the spatial organization of the neurons in *C. elegans* while an electrophysiologist might be interested in how signals generated in one neuron or population of neurons influence the electrical behavior of neurons elsewhere in the worm's nervous system. To understand how the worm controls its behavior requires one to combine these kinds of knowledge. Interfield integration (carried out either during a collaboration between researchers or after the hybridization of a single researcher into a multifield specialist) is required in biological sciences in part because of the diversity of constraints on any acceptable description of a mechanism and also due to the fact that researchers in different fields often study different kinds of constraints.

A second reason to expect integration to play such an important role in biology is that explanations in biology typically span multiple levels of mechanisms, linking facts about populations to facts about organisms, to facts about the cells and molecules within them. Consider the sense of touch in *C. elegans* again. One might begin by characterizing how the worm responds to different kinds of touch applied to different parts of its body. One might then look within the worm to find neural circuits that are especially dedicated to the task of carrying the information required to mediate these characteristic changes in behavior. One might then take an interest in the neurotransmitters that mediate communication among these cells or, perhaps, in the structure of the channels on the worm's membrane that respond directly to touch. Still other biologists might be interested in how the amino acids that compose the channels on the worm's surface fold into a conformation that allows them to function as a transducer for information about touch. Biology provides such fertile ground for interfield research because its subject matter is multilevel, and different fields are often identified with research at different levels.

Finally, biology requires interfield integration in part because it frequently deals with multiple interwoven temporal scales at once. One biologist might be primarily interested in how the touch system in *C. elegans* develops. Another might be primarily interested in the genes by which it is passed from one generation to the next. Another might be interested in how variation can arise in the system and how such variation contributes to the ability of adapted variants to evolve. One lesson of the last hundred years of biology is that the goings-on at these different time scales often cannot be understood independently of one another. Some emphasize, for example, that one cannot understand the evolution of creatures without understanding how genetic novelty influences developmental processes or how developmental growth fields constrain the space of possible variants. One cannot fully understand the organization of physiological systems without seeing them, in some sense, as having arisen as satisfying and heavily constrained adaptations in the face of environmental challenges and developmental patterns. Interfield perspectives are required to discover these connections.

The integration of biology is forged by building mechanism schemas that span many different levels, bridge across many different time scales, and that satisfy evidential constraints from many areas of biology (chemistry and physics too). From the perspective of a given phenomenon, one can look down to the entities and activities composing it. One can look up to the higher-level mechanism of which it is a component. One can look back to the mechanisms that came before it or by which it developed. One can look forward to what comes after it. One can look around to see the even wider context within which it operates. The adequate explanation of many biological phenomena requires describing a temporally extended and multilevel mechanism. This is why many fields, working at multiple levels, often must integrate their work in the discovery of mechanisms.

In what follows, we discuss exemplars of four different kinds of interfield integration, corresponding to the different reasons we have for expecting integrative research to play such a fundamental role in biological sciences. First, different fields might integrate their findings about different aspects of a mechanism, understood at a particular level. Such simple mechanistic integration arises when two or more fields investigate different stages in a mechanism, or different entities in a mechanism, or different aspects of the mechanism's organization. Their results are assembled, like pieces of a puzzle, to fill in the details of the mechanism and its organization. Different tools and perspectives are required to understand a mechanism made of different kinds of parts, activities, and forms of organization.

Second, different fields might integrate their findings across levels, either by looking up to see how a phenomenon is integrated within higher-level mechanisms or by looking down to see how a phenomenon is integrated with lower-level mechanisms. Such interlevel integration links part to whole through a nested hierarchy of mechanisms within mechanisms. In this special case different fields investigate phenomena that are related as part to whole. One field studies populations, another studies organisms, and another studies genes. One studies behaving organisms, one studies brain circuits, and one studies neurotransmitters or protein chemistry. The biological world is a world of mechanisms within mechanisms, often with behaviors and components that have been frozen into place through evolution to exhibit more or less stable and predictable behavior at many levels. The integrated schemas of contemporary biology reflect a world rich in structure and pattern at different spatial and temporal scales. They do not reflect a reductionist world in which everything is explicable in terms of the explanatory store of physics.

The third form of integration, intertemporal integration, is crucial in many areas of biology to understand how different processes or behaviors are related to one another in the temporal organization of a system. Depending on the particular features of the organization of the mechanism in question, it may be more appropriate to talk of intertemporal or interlevel integration, and in many cases it will be appropriate to talk in terms of both. Often the mechanistic scaffolding for integrating fields is organized in terms of relationships among temporal processes or stages; we call this *intertemporal integration*. In some simple cases, such as the relationship between Mendelian genetics and molecular biology, our third example, different fields contribute to an understanding of the mechanism of heredity at different stages. In more complex intertemporal cases, such as the evolutionary synthesis, our fourth example, different fields contribute to an understanding of different mechanisms that act at multiple levels contemporaneously and over different time scales to continually contribute to the phenomenon.

SIMPLE MECHANISTIC INTEGRATION: DISCOVERING THE MECHANISM OF PROTEIN SYNTHESIS

Most mechanisms in the biological world are heterogeneous. They are composed of many different kinds of components. And the components interact via diverse activities. Different parts of the mechanism are organized differently. The resources of many fields of science are required to investigate and schematize such a multifaceted thing.

Different fields of biology have different experimental techniques and different accepted protocols that allow them to ask particular kinds of questions (and not others). Different fields might have the tools to understand different stages of the same mechanism, or different entities and activities within it, or different aspects of their organization together. One field might work on the beginning of the mechanism, while the other starts from the end. One might have the techniques to find the chemical or physical nature of a component, while the other attends to the overall organization and supplies the roles played by those components. One might focus on causes early in the mechanism, while the other pays attention to the effects downstream. This simple mechanistic form of integration is nicely exemplified by our extended example of the discovery of the mechanism of protein synthesis.

In the 1950s and 1960s molecular biologists and biochemists worked from different ends of the mechanism of the protein synthesis and ultimately met in the middle. Molecular biologists, such as Watson and Crick, were primarily interested in understanding how the flow of genetic information, namely how the linear pattern in the sequence of bases in DNA, produces the base sequence in RNA and the linear order of the amino acids in proteins. Biochemists were primarily interested in the flow of matter (amino acids) and energy (required for chemical bond formation); they wanted to know how single amino acids are strung together into long chains (without regard to the particular sequence). Molecular biologists asked what role genes play in synthesizing proteins. They investigated the role of the newly discovered (1953) double helix structure of DNA and hypothesized what role the genes, viewed as sequences of bases along segments of the DNA double helix, could play in ordering amino acids in proteins. Biochemists, on the other hand, focused on the chemical reaction by which peptide bonds form between amino acids, which they then viewed as the primary activity in the mechanism of protein synthesis.

Despite these differences scientists in both fields aimed at discovering the mechanism of protein synthesis. Physician-turned-biochemist Paul Zamecnik (1912–2009) described these two fields as excavators boring a tunnel from two sides. Their work met in the middle: at RNA. To complete the connection, the fields had to sort out three different kinds of RNA and to describe the roles of these three kinds of RNA in the protein synthesis mechanism.

As we have discussed in previous chapters, molecular biologists investigated numerous hypotheses about the players that filled the role of the template between the sequence of bases in the DNA and the ordering of amino acids in proteins. By 1961 they had discovered messenger RNA, which is complementarily copied along the bases of DNA and then attaches to a ribosome (a large particle composed of two segments of RNA and several previously synthesized structural proteins). The order of the bases in messenger RNA produces the order of the amino acids in the protein. But the amino acids do not attach directly to the messenger RNA. The molecular biologist Francis Crick predicted the existence of what he called adaptor RNA molecules, which are today called transfer RNAs. Twenty different adaptors, he proposed, would attach to the twenty different amino acids. Each adaptor would have a specific sequence of three bases that would complementarily bond to a sequence of three bases on the template RNA, placing the amino acids in a position to bond to the one next to it in a growing protein chain.

As a molecular biologist, Crick was concerned with genetic mechanisms. He explicitly used tools from structural chemistry. He focused on the structures of the macromolecular nucleic acids and on how slight charges in those structures enabled weak but specific bonds to form between nucleic acid bases.

Meanwhile biochemists were decomposing protein synthesis systems via centrifugation and investigating which components were needed to induce peptide bonds to form between amino acids. Zamecnik and Mahlon Hoagland (1921–2009) investigated the energy requirements for producing the strong peptide bonds between amino acids. They worked in an in vitro system in which they tried to induce the individual amino acid molecules to bond into polypeptide chains, one step in the synthesis of a protein. They labeled the amino acids with radiotracers so that they could follow them through the mechanism. To Hoagland's surprise, activated (energetic) amino acids were attached to RNAs prior to their incorporation into a growing chain of amino acids. In the early 1950s, prior to their interactions with molecular biologists, biochemists had not assumed that DNA or RNA played any role in the synthesis of proteins. Their schema is a typical chemical reaction diagram, in which reactants plus energy yield the product.

$$aa_1 + pppA + E_1 \rightleftharpoons aa.pA_1 \cdot E_1 + pp$$

The amino acid (aa) combines with ATP (the energy source, adenosine triphosphate, here pppA), and an enzyme (E) catalyzes the reaction. Two phosphates (pp) are released and the other product of the reaction is an activated amino acid ($aa.pA_1$). Hoagland said that this reaction provided "a mechanism for activation of amino acids" (Hoagland, 1955, p. 288). Thus, Hoagland found an intermediate in the reaction between free amino acids and the bound ones in a polypeptide (a string of covalently bonded amino acids). This activated amino acid was a high-energy intermediate in the reaction, which biochemists expected as they investigated the flow of matter and energy in a chemical reaction. However, quite unexpectedly, Hoagland found that this activated amino acid was attached to a bit of RNA. Proteins do not contain RNA as a component. In protein synthesis, RNA did not appear to be an energy source or a catalyzing enzyme. It did not fit into a biochemist's mechanism schema.

What role could RNA be playing in the protein synthesis mechanism? In 1956 the molecular biologist James Watson visited Hoagland, the biochemist. Watson informed him of Crick's adaptor hypothesis. Additional work proved the existence of transfer RNAs (specific to their respective amino acids), which attach to the amino acids, transport them to the ribosomes, and selectively bond to complementary sequences along the messenger RNA. In doing so, they arrange the amino acids in the proper order for the peptide bond to form with the one next to it in the growing protein chain.

This *interfield intralevel* work (what we call *simple mechanistic integration*) resulted in an understanding of the roles of the three different types of RNA in the middle of the protein synthesis mechanism: ribosomes are the site of synthesis, messenger RNA is the template, and transfer RNA conveys the amino acids to their proper place on the template. In this case the two fields had different fundamental questions. Molecular biologists were interested in a mechanism that would echo or preserve the linear order of bases in DNA in the order of the amino acids in the corresponding proteins. Biochemists were interested in the mechanisms of incorporation; they wanted to learn how individual amino acids bond in polypeptide chains. These are two questions that any adequate schema for the mechanism of protein synthesis must be able to answer. Yet answering them required different tools. The molecular biologist's project required reasoning about the flow of information and about the role of weak polar bonds that form and break easily among components of nucleic acids (DNA and RNA). The biochemists used cell fractionation and in vitro experimental systems; they investigated the roles of different cellular constituents and the energy requirements for forming strong covalent bonds. Neither field, taken in isolation, could tell the complete story of protein synthesis; the stories had to be integrated.

We might just as easily have chosen any number of examples from the history of biology. Given the heterogeneity of biological mechanisms, the limited techniques of a single field are rarely sufficient to bring the whole mechanism into view. Indeed, as biology continues to develop, one finds that individual researchers boldly cross borders between fields in search of mechanisms. Sometimes, new integrative fields (developmental neurobiology, integrative biology) come into existence.

INTERLEVEL INTEGRATION: DISCOVERING THE MECHANISMS OF LEARNING

Most mechanisms in the biological world can be described as spanning multiple mechanism levels. They are mechanisms composed of mechanisms, which are themselves composed of mechanisms, and so on. Herbert Simon (1916–2001) called mechanisms that have this kind of structure *nearly decomposable systems*. He argued on theoretical grounds that systems that have arisen through natural selection are most likely nearly decomposable. The spirit of Simon's argument is this: creatures that can adjust one component of a mechanism independently of others (either within an individual or across generations) are more likely to survive environmental challenges without a global collapse. If so, one should expect biological systems to be composed of more or less independent mechanisms assembled into higher-level mechanisms, and so on.

An alternative route to Simon's theoretical conclusion is to open any standard introductory biology textbook and look at some of the explanations. One quickly discovers that explanations in biology typically bounce back and forth among a number of different levels in a hierarchy of mechanisms. One learns that many of the great achievements in the history of biology involve bridging different levels of mechanisms. Sometimes scientists *look down* to a deeper level of the hierarchy. Hodgkin and Huxley (see Chapter 3) explained the electrophysiological properties of cells in terms of the underlying flux of sodium and potassium ions across the cell membrane. Molecular biologists described the genetic code that underlies heredity in whole organisms. Other times, scientists look up in the hierarchy to find a higher-level mechanism within which the phenomenon is situated. H.B.D. Kettlewell (1907–1979) had a hypothesis that dark-colored moths would escape predation more readily than light-colored moths in the soot-stained woods around Birmingham, England. William Harvey posited that the heart might function to propel blood in a circuit through the arteries and veins (as we discussed in Chapter 7).

Both the downward- and upward-looking forms of interlevel integration have contributed essentially to recent research on the neural mechanisms of learning and memory. A dominant theme throughout the history of this research area has been that learning and memory are ultimately grounded in synaptic plasticity: use-dependent changes in the efficacy or efficiency with which one neuron influences the behavior of another across the synapse (the gap) between them. This idea survives in contemporary neuroscience. Researchers now recognize many distinct kinds of synaptic plasticity. But the first kind of synaptic plasticity to fully capture the scientific imagination is known as Long-Term Potentiation (or LTP). As its name suggests, LTP is a long-lasting and use-dependent strengthening of the synapse. It can be artificially induced in cells by delivering a rapid and repeated electrical stimulus (known as a tetanus), but it is also known to be induced during some forms of learning, to be necessary for some forms of learning, and (in some cases) sufficient in the circumstances to induce changes in the behavior of whole organisms. Thousands of researchers are currently dedicated to working out aspects of LTP, its mechanisms, and its relation to learning and memory.

LTP became part of the neuroscientific literature as a result of three papers written by T. V. P. Bliss, Terje Lømo, and A. R. Gardner-Medwin and published in 1973. It is hard to say whether these authors were the first to observe the phenomenon that would come to be called LTP. Some researchers had described a kind of "kindling" (or amplification) in hippocampal circuits that could, if unchecked, lead to epileptic seizures. Other electrophysiologists in the 1950s and 1960s noted that hippocampal synapses change when they are rapidly stimulated, but these notes are typically buried deep in lengthy articles on diverse aspects of hippocampal electrophysiology. Per Andersen, in whose lab Bliss, Lømo, and Gardner-Medwin worked, reports that most electrophysiologists of the day understood the effect simply because they could use it to "reawaken" a cell that was beginning to deliver a very weak recording signal.

What changed in 1973? No doubt part of the answer concerns the fact that Bliss, Lømo, and Gardner-Medwin characterized the phenomenon explicitly in terms of specific electrophysiological changes and demonstrated that the phenomenon could be reliably produced time and again. And no doubt part of the answer concerns the fact that they demonstrate that the phenomenon is not an artifact of anesthesia by producing it in unanaesthetized rabbits. Yet another crucial part of the answer concerns how Bliss, Lømo, and Gardner-Medwin conceptualized LTP as part of a multilevel mechanism, one that reaches up to the phenomenon of learning and down into the molecular mechanisms for the electrophysiological and chemical activities in cells.

Had LTP remained nothing more than electrophysiological phenomenon, it could not have generated the overwhelming attention it has received since 1973. The connection with learning and memory, in the domain of experimental psychology, made it exciting. The authors point out that synaptic plasticity has long been hypothesized to underlie forms of learning and memory. They note that LTP lasts for a very long time, as one might require of a memory trace. They also attempt to bridge the connection between the electrophysiological phenomenon of LTP and the psychological phenomenon of learning and memory through an intermediate, gross-anatomical level of the hippocampus, an organ that was coming to be associated with different forms of learning and memory. Wilder Penfield reported in the 1950s that brain surgery patients reported apparent memories when he stimulated areas around the hippocampus with his electrode. And scientists were increasingly become aware of findings that patients with lesions to the hippocampus suffer debilitating amnesias. Bliss, Lømo, and Gardner-Medwin discuss their results explicitly in these terms, suggesting that the otherwise mundane electrophysiological phenomenon of LTP might rise above the rabble through its upward-looking connections with neuropsychology and experimental psychology.

Yet the framework they sketch around LTP also has a downward-looking component. Bliss, Lømo, and Gardner-Medwin characterize the LTP phenomenon with sufficient clarity that it could guide the search for the molecular mechanisms that underlie long-term potentiation. They dedicate much of their discussion to possible molecular mechanisms that might underlie LTP. The search for those mechanisms continues to the present.

The LTP phenomenon provides some guidance in the search for a lower-level mechanism. LTP is induced only when both the presynaptic neuron and the postsynaptic neuron are simultaneously active. These are the precipitating conditions for LTP. A sketch of the mechanism that might account for this fact about LTP is shown in Figure 10.1. When the presynaptic neuron is active, it releases glutamate (GLU). This glutamate binds to N-methyl-D-aspartate (NMDA) receptors on the postsynaptic cell. The NMDA receptors change their conformation, exposing a pore through the cell membrane. If the postsynaptic cell is inactive, the channel remains blocked by large Mg^{2+} ions. But if the postsynaptic cell is depolarized, these Mg^{2+} ions float out of the channel, allowing Ca^{2+} to diffuse into the cell. The rising intracellular Ca^{2+} concentration sets in motion a long chain of biochemical activities terminating in the question marks of Figure 10.1. Those question marks stand for a large space of possible mechanisms. Some posit that there are rapid changes to the number of active receptors on the post-synaptic cell, perhaps by activating dormant receptors. There are also thought to be long-term changes that require the synthesis of proteins in the postsynaptic cell body. These proteins are then used to alter the structure of the dendritic spines at that synapse. Some suspect that there is also a presynaptic component of the LTP mechanism, whereby, for example, the presynaptic cell releases more glutamate.

Figure 10.2 illustrates the effect of zooming out to take in the whole scaffold that integrates LTP into multiple levels in the mechanisms of learning and memory. The description of this mechanism includes mice learning to navigate through mazes, hippocampi generating and storing spatial maps, synapses inducing LTP, and macromolecules (like the NMDA receptor) binding and bending. The levels in this hierarchy stand in part-whole relations to one another, and the lower-level entities and activities are components of the higher-level mechanism. The binding of glutamate to the NMDA receptor is a lower-level activity in

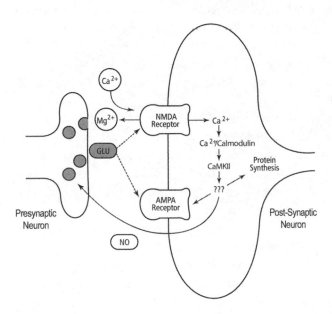

Figure 10.1 A cartoon sketch of a possible LTP mechanism as understood in the early 1990s.

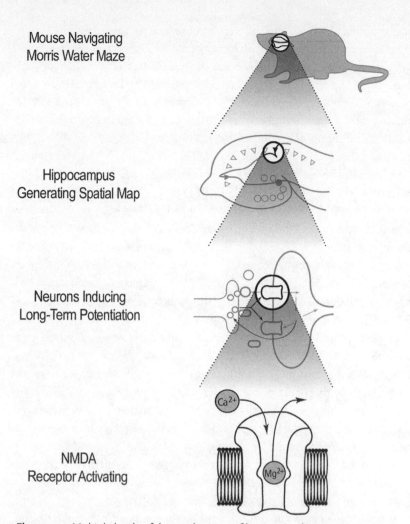

Mouse Navigating
Morris Water Maze

Hippocampus
Generating Spatial Map

Neurons Inducing
Long-Term Potentiation

NMDA
Receptor Activating

Figure 10.2 Multiple levels of the mechanism of learning and memory.

the mechanism of LTP, and LTP is thought to be a lower-level activity in spatial map formation, which, of course, is thought to be an activity in the mechanisms of learning and memory.

LTP came into its own as a research topic because it became possible to see it as integrated into a multilevel mechanism. This mechanism reached upward to the phenomena of learning and memory studied by experimental psychologists, and it reached downward into the molecular mechanisms that underlie the electrophysiological and chemical properties of neurons. This integration transformed LTP from a curiosity, a pathology, and an experimentalist's trick into a phenomenon in its own right. At the same time, this integrative project sketched a framework around LTP that could accommodate the contributions of diverse researchers. The lesson is general: Biologists study a world of mechanisms within mechanisms. Interlevel integration is the scientific process of learning how these different levels of mechanisms fit together into a single functioning mechanism. Multilevel integration is required if one is to grasp the richness and complexity of a world of evolved and evolving creatures.

SEQUENTIAL INTERTEMPORAL INTEGRATION: SEQUENTIAL RELATIONS AMONG MENDELIAN GENETIC AND MOLECULAR BIOLOGICAL MECHANISMS

Sometimes the idea of levels of mechanisms is less useful for understanding how diverse areas of biology are integrated than is the idea that workers in different fields describe phenomena as occurring at different times and across different time scales. Different fields might work on different stages of a mechanism operating sequentially across generations. Different fields might work on different processes that contribute contemporaneously to the production of a given effect yet also operate again and again across a longer time scale. We discuss these two kinds of intertemporal integration in turn.

In the first case two or more fields are integrated by the fact that they work on submechanisms that occupy different stages of a more inclusive mechanism. In serial mechanisms one field might work on a mechanism at an earlier stage while another field focuses its energies on a submechanism at a later stage. (In mechanisms with feedback, the language of earlier and later becomes strained, and it is more appropriate simply to say that they work on different stages in such a mechanism.) As we will see, the science of heredity involved the piecemeal completion of a sequential mechanism sketch, with researchers in different fields struggling to understand different submechanisms in that sketch. In the second case evolutionary biology requires researchers to span multiple different time scales, from molecular events lasting seconds, to occurrences within the lifespan of individual organisms, to population effects lasting many years, to branching species formation occurring over eons. In this case a more complex form of temporal integration is required. Consider these examples in turn.

One puzzling phenomenon of heredity is that offspring only *partially* resemble both their parents. Perhaps the daughter has her father's green eyes and red hair, but she also has her mother's strong white teeth. The central problem for the field of genetics is to understand these hereditary relations. Two different subfields of genetics addressed different aspects of this problem and, in the end, produced an integrated schema of the mechanisms of heredity. Researchers in classical, Mendelian genetics (1900 to 1930) sought to explain the regularities approximately captured by Mendel's laws of segregation and independent assortment. Researchers in early molecular biology (1953 to about 1970), in contrast, sought to explain details about how genes are copied, about how mutations arise, and about how genes produce hereditary characters. In short the two fields were addressing different stages of a temporally extended mechanism. Both fields contributed to producing an integrated picture of the mechanism of heredity.

The fields of Mendelian genetics and molecular biology discovered separate but serially integrated mechanisms. These hereditary mechanisms have different working entities and the mechanisms operate at different times in an *integrated temporal series*. The working entities of the mechanisms of Mendelian heredity are chromosomes. The regularities captured in Mendel's laws of segregation and independent assortment are implemented by the chromosomal mechanisms during meiosis (the formation of germ cells). The working entities of numerous mechanisms of the molecular biology of the gene are larger and smaller segments of DNA plus related molecules. Molecular DNA mechanisms, discovered in the early days of molecular biology (1950s to the 1970s), filled black boxes at stages before and after the chromosomal mechanisms that the Mendelian/cytological techniques of the 1930s and 1940s did not illuminate.

Figure 10.3 graphically depicts the serially connected submechanisms within the mechanism of heredity. The diagram depicts the formation and behavior of germ cells, leading to fertilization and then the development of the fertilized egg.

Begin with the formation of germ cells (sperm and egg, called gametes) in the first stage. Meiosis is the chromosomal mechanism of gamete formation. The chromosomes duplicate. Geneticists assumed that genes are parts of the chromosomes and that they replicate during the duplication of the chromosomes, but the mechanism for this stage was, around 1930, left as a black box for future work. In the second stage of this mechanism, parts of homologous chromosomes sometimes cross over, switching a piece of the mother-donated chromosome with a piece of the father-donated chromosome. Again, classical geneticists as of 1930 left the mechanism underlying this chromosomal crossing over unspecified and black boxed. Following the arrows in Figure 10.3 to the next stage, we see next the chromosome pairs lining up. The chromosomes from the mother and the father assort independently along this line. Hence, in the following stage, when they separate (technically, segregate), those from the mother and from the father are mixed together more or less randomly. As the gametes form they get one or the other of a homologous chromosome pair, but not both. When mating occurs the fertilized egg has a complete complement of chromosomes

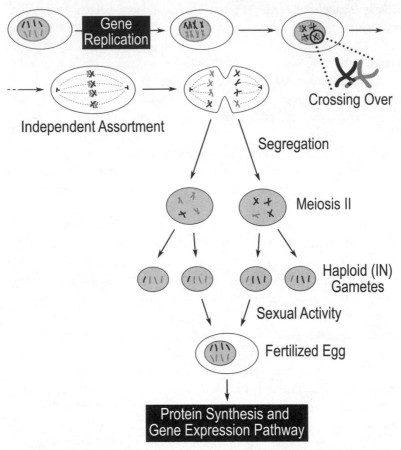

Figure 10.3 The serially connected submechanisms of the mechanism of heredity.

(for humans, that's 23 pairs). As embryological development occurs, genes (later found to be segments of DNA that are parts of chromosomes) are expressed and play a role in the formation of observable characters, such as hair and eye color. Again, in the 1930s, the mechanisms operating between genes and characters were left as black boxes.

The chromosomal mechanisms of crossing over, independent assortment, and separation implement the regularities stated in Mendel's laws. Using the technique of crossbreeding of variants of plants and animals, classical geneticists of the early twentieth century noted the distributions of the variant phenotypic characters through several generations. They then made inferences about hypothetical genes associated with those characters. Most genes come in pairs, called alleles, one from the mother, the other from the father. Mendel's first law of segregation states that in the formation of gametes, the alleles of a gene segregate so that each germ cell gets one or the other of the two alleles of a gene, but not both (and no blend form). Mendel's second law of independent assortment states that genes in different linkage groups assort independently of each other. Furthermore, linkage is not complete; sometimes alleles cross over, the pieces in the two linked groups exchange places. All this shuffling ensures that children are not exactly like their parents but have a mix of characters from both.

As the diagram shows, the chromosomes are the primary working entities in the mechanisms that explain the regularities of Mendel's laws. In 1915 T. H. Morgan and his colleagues called attention to this relationship with the title of their book, *The Mechanism of Mendelian Heredity*. They stated: "The chromosomes furnish exactly the kind of mechanism that the Mendelian laws call for. . . . Moreover, as biologists, we are interested in heredity not primarily as a mathematical formulation but rather as a problem concerning the cell, the egg, and the sperm" (Morgan et al. 1915, p. ix). Discovering these mechanisms required looking up. The genes are parts of chromosomes that are parts of germ cells. During the formation of sperm and eggs and their fusion during fertilization, the genes ride along with the chromosomes, getting assorted and segregated and recombined as those higher-level mechanisms operate.

By the 1930s geneticists and cytologists had a number of outstanding and somewhat pressing problems. They did not know the chemical nature of genes. They did not know how genes replicate. They did not know how genes produce phenotypic characters. They did not know how genes mutate and faithfully reproduce those mutations. Molecular biologists looked down inside the chromosomes to fill those black boxes (see black boxes of Figure 10.3). They directed their attention to various stages identified by the classical geneticists.

Watson and Crick's 1953 discovery of the double helix structure of DNA marks the beginning of the modern field of molecular biology. The new field drew on work in several other fields: X-ray crystallography, for determining the structures of macromolecules; structural chemistry, especially Linus Pauling's work on weak forms of chemical bonding, such as hydrogen bonding; and to a lesser extent biochemistry, for its study of the chemical analyses of proteins and nucleic acids, as well as energy requirements of strong covalent bonding.

In its early days the field of molecular biology did not tackle the molecular details of chromosome pairing, crossing over, and separation. Instead, the molecular biologists centered their work on the key unsolved problems about genes. That is, they began filling the black boxes at different stages of the hereditary mechanism identified by Mendelian geneticists that could be addressed with more easily manipulated microbes. For their model experimental systems they chose viruses and bacteria, which did not have the same kind of organized chromosomes as did eukaryotes, such as the fruit fly (*Drosophila melanogaster*), a favorite model organism of Mendelian geneticists.

As Watson and Crick noted, the structure of DNA immediately suggested a copying mechanism for the genetic material. The DNA double helix opens to serve as two templates for the assembly of like complements into matching strands. The structure also immediately suggests a means by which at least one kind of mutation can be produced. Point mutation, a copy error, occurs during DNA replication when a base substitution departs from the usual A-T, G-C base pairing, for example, if an A-C pairing occurs. A single base change can result in subsequent effects during protein synthesis, which can produce a different phenotypic character.

Interestingly, just as genes are not the working entities of the chromosomal mechanisms, genes (functional sequences of bases along one strand of the DNA helix) are not the working entities of either gene replication or gene mutation. The entire DNA double helix molecule replicates. Moving down in size, the working entity for the formation of a point mutation is the single base. The genes are the working entities only in the mechanisms of gene expression, namely protein synthesis and its regulation (and that view has been qualified by later discoveries, mostly in eukaryotes, about intervening sequences among functional gene segments and various kinds of reshuffling of those functional segments during gene expression). The different fields thus worked on mechanisms at various mechanism levels, with working entities at different size levels from phenotypic characters in organisms, to germ cells, to chromosomes, to the entire DNA double helix, to segments of that helix. The scaffolding mechanism of heredity served to integrate submechanisms at these different levels into a serial temporal sequence.

In sum, we see that early molecular biologists worked to fill in the picture of the temporally extended mechanism of heredity. They discovered different molecular level mechanisms operating both before and after the Mendelian/ cytological ones for assortment and segregation. They focused on the nature of the gene rather than on the molecular details about chromosomal mechanics. Their work filled the black boxes with the mechanisms of gene replication, mutation, and some mechanisms of gene expression (although the many different mechanisms of gene regulation are still lively areas of research). The diagram of Figure 10.3 depicts the mechanism of heredity, which scaffolds the submechanisms that operate in a serial, temporal order, discovered by different fields spanning much of the twentieth century.

This episode required the integration of two different fields because neither field had the tools required to solve the problems addressed in the neighboring field. Classical geneticists could rely on breeding experiments to reveal patterns in the inheritance of phenotypic characters, but they did not have the equipment (either technologically or conceptually) to address questions about, for example, the makeup of chromosomes or the means by which such chromosomes replicate. They could identify features of the way hypothetical genes behaved, but they relied on cytologists to identify the subcellular structures (the chromosomes) and their behavior that paralleled that of hypothesized genes (independently assorting and separating). The Mendelian/cytological mechanism schemas bottomed out well before all of the interesting and important questions could be answered. Molecular biologists had the tools to investigate macromolecules, but their behaviors take on significance only in the context of the higher-level mechanism of heredity. This mechanism has multiple sequential stages: molecular and chromosomal mechanisms at work in the cells of individual organisms fit into a larger temporal context of hereditary relations between parents and their offspring. Mendelian genetics and molecular biology were integrated through the elaboration of this multistage mechanism, with its temporal serial stages studied by the different fields.

CONTINUOUS INTERTEMPORAL INTERLEVEL INTEGRATION: THE EVOLUTIONARY SYNTHESIS

Charles Darwin (1809–1882) took as his central intellectual problem the task of explaining the presence and prevalence of adaptive traits in populations. Darwin sketched a solution to this problem, and this sketch has been elaborated and revised repeatedly in the 150 years since Darwin's 1859 version. As the title of his most famous book suggests, Darwin also intended to explain the origin of new species and the branching shape of the tree of life. This aspect of Darwin's theory has also been revised substantially in light of discoveries concerning the mechanisms that isolate one species from another.

Darwin's genius was to put forward a bold mechanism sketch that could possibly explain these phenomena and, perhaps just as importantly, that raised questions that had to be answered by many different fields across biology. By now we should not be surprised to find that different aspects of this modular mechanism sketch have had to be revised and reconceived in light of these findings from different fields of biology. The period of the 1920s and 1930s marked one of the most important periods in the formation of the modern view of the mechanisms producing evolution. The *evolutionary synthesis*, as it is called, is sometimes characterized merely as integrating new findings from Mendelian genetics with Darwinian natural selection. Although that is a key part of what the synthesis accomplished, it did more too, adding a new level of speciation mechanisms.

In his 1937 book, *Genetics and the Origin of Species*, Theodosius Dobzhansky (1900–1975), one of the chief architects of the synthesis, argued for a three-level perspective on the mechanisms of producing the diversity of life on earth. The first level includes genetic and chromosomal changes that produce variations; the second level includes the mechanisms that mold the structures of populations, including natural selection; and finally the third level concerns the fixation of diversity via isolating mechanisms to yield new reproductively isolated species (Dobzhansky 1937, pp. 12–13). The mechanisms at each of these levels operate continuously; over time they may give rise to adapted traits and new species (or else extinction). Thus, following Dobzhansky, we see the mechanism of evolution as a multilevel, temporally extended mechanism with many simultaneous instances of operation (runs) of its submechanisms: in evolution variations are continuously produced; the mechanism of natural selection (and other population changing processes such as migration) continuously operate; and, under the appropriate conditions, isolating mechanisms arise, leading to the production of new reproductively isolated gene pools, i.e., new species. We call this *continuous intertemporal interlevel integration*. Now let's discuss each of these levels and the mechanisms operating at each.

As we described in the preceding section, the rise of Mendelian genetics in the early twentieth century added knowledge about kinds of mutations and chromosomal changes and the Mendelian/cytological mechanisms by which they are inherited. By the 1930s Darwin's speculations about heredity were replaced. Mendelian genetics and the study of larger scale chromosomal changes, Dobzhansky emphasized, provided the findings about raw materials for evolutionary change. Dobzhansky himself investigated the presence and prevalence of chromosomal changes in wild populations of fruit flies (the famous model organism *Drosophila*) and established their importance in distinguishing species.

Later, molecular biology would add even more to this first level of variant production by finding the mechanisms by which DNA mutations arise, such as during imperfect copying of the DNA double helix, and the mechanisms by which genes (viewed as segments of DNA) are inherited. This achievement has prompted some to suggest enthusiastically that evolution might be understood entirely from the "genes-eye-view": that the story of the origin of species might in fact be a story only about shifting ratios among genes in populations over time. In fact in the 1970s Richard Dawkins (b. 1941) developed such a boldly reductionist view of selfish genes struggling to endlessly reproduce themselves.

From the mechanistic perspective, such bold reductionism is an unnecessarily impoverished vantage point from which to think about evolution. It is, we admit, fascinating to see just how far one can get in thinking about evolution from the gene's perspective. But one should not confuse a clever formulation of a theory with the one true formulation of the theory. Mechanisms can frequently be described in many different ways, with different items in the foreground and in the background, accentuating different pieces of the mechanism depending on one's practical purposes. Natural selection is a complex mechanism with submechanisms at different stages of a temporally extended mechanism and operating at different levels from molecules to species, as Dobzhansky's original vision made clear.

The second level of evolutionary mechanisms is where natural selection operates. As Dobzhansky said: "No coherent attempts to account for the origin of adaptations other than the theory of natural selection and the theory of the inheritance of acquired characteristics have ever been proposed" (Dobzhansky 1937, p. 150). As we discussed in previous chapters, instructive theories had been proposed repeatedly in the history of biology, with Lamarck's theory the most famous. However, Mendelian genetics and cytological work on the isolation of the germ cells early in embryological development ruled out such instruction. That work showed the difficulty of getting instructions from the environment back into the genetic material to make an adaptive mutant that is passed to the next generation. By the time of the synthesis, the only viable remaining scientific explanation of the origin of adaptive traits was (and still is) Darwin's mechanism of natural selection. The integration of the Mendelian/cytological hereditary mechanisms and Darwinian natural selection explains the presence and prevalence of adapted traits in populations.

The mechanism of natural selection itself operates on two levels, the level of the individual organism's joint activities with an environmental challenge and the level of the population, in which some variants do better than others in the face of that challenge. Over an extended period of time, selection occurs when some variants survive and reproduce in greater numbers than others because of the success of their activities in responding to that environmental challenge. The adapted trait is like a tool crafted to fit the need created by the environmental challenge; such a trait is engineeringly fit. This seemingly designed trait, however, has arisen by a spontaneous change (see the mechanisms for variant production on the first level) and just happens to be well designed, or at least good enough, to give the organism possessing it an advantage (slight or decisive) over others without that variant trait. Selection is comparative: at least one other variant must have (or have had in the past) better activities in that environment than another. So, as a population of variants acts and interacts in a common challenging environment and when those with one trait work better than others, then those adapted traits tend to increase in frequency in the population.

A very wide range of different kinds of traits and environmental challenges can play roles in the mechanism of natural selection. Finches' beaks allow them to eat some types of food and not others. The jackrabbit's large ears allow it to dissipate heat in a hot climate. Some orchids have a structure that mimics the shape of a female insect; when the male insect attempts to mate with it, the orchid is pollinated. (Note that in the orchid case, the benefit is for reproduction, not the survival of the parental orchid plant itself.) Barring catastrophes (and even minor mishaps), the most engineeringly fit organisms tend to leave more offspring that also fare better in the same environmental challenge their parents faced. These multilevel mechanisms operate over temporally extended periods, thereby producing population changes.

As an example of the various levels and temporal stages in the mechanism of natural selection, consider the marine biologist Robin Seeley's study in the 1980s of the rapid change in the shell structure of the intertidal snail Littorina obtusata, commonly called a yellow periwinkle. Between 1871 and 1984 the shell structure in some L. obtusata snail populations in waters off northern New England changed dramatically. High-spired shells predominated in 1871. By 1984 low-spired shells predominated. Seeley argued that natural selection is the mechanism that produced the rapid change in the prevalence of low-spired shells in the periwinkle population.

Seeley described the phenomenon to be explained. Snail shells collected prior to 1900 in northern New England are high spired with thin walls, but shells from L. obtusata populations in most areas of northern New England by the 1980s were low spired with thick walls. The change in shell structure coincided with the expansion of the range of Carcinus maenas, a crab that preys on L. obtusata. Prior to 1900 this green crab was not found in waters north of Cape Cod. This invasive species is now found in most, but not all, of the waters where the yellow periwinkle is found. Although the low-spired shell predominates in most of L. obtusata's range, individuals with high-spired shells can still be found in waters where C. maenas is rare.

A series of laboratory and field studies provided evidence about the organism-level activities that compose this predator-prey mechanism and illustrate the many kinds of techniques that evolutionary biologists employ. High-spired L. obtusata shells are more vulnerable to green-crab predation than low-spired shells. In laboratory tests, when green crabs were offered a high or low spired snail, only 12% of the low-spired snails but 100% of the high-spired snails were successfully attacked by the crabs. The crabs attack the snails with their claws, crushing the snails' high-spired shells. Low-spired shells have overlapping whorls. Because shell thickness increases with each successive whorl, increased overlap results in the thin shell whorls of the juvenile snail being enclosed in the thicker whorls of the adult. In high-spired shells, these thin (and thus more vulnerable) whorls are not encased and are exposed to crushing predators.

We thus see how the differential effects of the variant properties act differently in the crushing mechanism to produce the increase in the prevalence of the adapted property in the population subject to predation. In this study, units, activities, and mechanisms at various levels—genes for the variant snail-shell shapes, variant phenotypic properties, organisms, the environmental predatory challenge, and the change in a subpopulation—all are integrated in an account of the mechanism of natural selection. One could, we suppose, tell a story only about changing gene frequencies in different populations, but to do so is to black box the activities of the variant organisms during the environmental challenge, the crucial difference-making step of the natural selection mechanism. In the case of the snails, to understand how selection changes the gene frequencies, one has to understand how shell shapes influence susceptibility to predation. In other cases one might need to appeal to levels above the single organism, to talk about sexual selection, kin selection, and perhaps selection of larger groups, such as colonies or tribes. The individual activities at the organismic level are one level, but to see selection operating one must also incorporate the population level: the relative frequency of predation among different shell-shapes. As a result of these organismic-environment activities, populations gradually and probabilistically change their composition over shorter or longer time scales. To understand the temporally extended changes at the population level, we must see how crucial difference-making activities at the level of organisms in a challenging environment lead to changes in populations over time.

Now that we have discussed mechanisms at the levels of variant production, joint organism-environment activities, and populational changes, we can discuss the additional level of mechanisms for producing new species. Darwin did not have a separate species level. The evolutionary synthesis added a new level and began the investigation of isolating mechanisms. Darwin argued that natural selection not only produced adaptations but also operated in the origin of new species. His view about this is now referred to as the *principle of divergence*. If organisms are competing for a shared resource, variant traits that let some organisms utilize different resources, not in such demand, will convey a selective advantage. Darwin suggested, for example, that if many carnivorous quadrupeds live in an area where they are competing for the same prey, then any variant that allowed the animal to feed on new kinds of prey, dead or alive, would convey an advantage. In the area where the most competition raged between those eating the same prey, more death of the carnivores would often ensue. Over many thousands of generations, Darwin claimed, such differences would, in some cases, be sufficient to drive a wedge into the group so as to separate it into two distinct species, utilizing different resources.

Dobzhansky and other population geneticists of the evolutionary synthesis argued that more was needed than Darwin's principle to prevent interbreeding (and thereby the continued mixing of the gene pool) to establish two distinct species. Dobzhansky added an additional species level and he coined the term *isolating mechanisms* to refer to mechanisms operating to maintain separate gene pools. One of the most widely accepted views of speciation was (and is) that of another architect of the synthesis, Ernst Mayr (1904–2005). If a geographic separation between two parts of a population occurs (thereby preventing their interbreeding), then different selection pressures on the different subsets of variants in the two groups over time likely give rise to new species (unless one dies out or they rejoin before sufficient time under different selection regimes has passed). Over time, biological isolating mechanisms arise, such as chromosomal pairing incompatibilities. (Some researchers claim to have discovered mechanisms for speciation without geographic separation, but debate on this is ongoing.) Work continues to find isolating mechanisms, with the search for speciation genes an active area of research.

In sum, the phenomena of adaptations and the branching tree of life are produced by a multilevel mechanism, with natural selection at its core. Multiple fields study the mechanisms of variant production, the way natural selection operates, and the isolating mechanisms giving rise to new species. Many kinds of entities and activities at different levels of mechanisms and across different time scales are integrated in the mechanism producing evolutionary change. Its stages usually operate continuously, over extended periods of evolutionary time. Scientists still debate whether additional mechanisms are needed to explain the origin of higher level taxa, such as genera, families, orders, phyla, and kingdoms as diverse as animals and plants, trees and dinosaurs, pines and *T. rex*. But that debate is beyond the scope of our discussion here.

CONCLUSION

The science of biology must be integrated because it deals with a domain of heterogeneous phenomena, because mechanisms span multiple levels, and because mechanisms often operate at and across different time scales. In simple mechanistic integration, fields work on different aspects of one and the same mechanism at a given level. In interlevel integration, researches form bridges that span the behaviors of mechanistic wholes and the behaviors of their parts. In sequential and continuous intertemporal cases, researchers have to learn to integrate phenomena at different times or operating at vastly different timescales.

The search for mechanisms in biology has produced a matrix of integrated biological knowledge, in which one understands the role of heredity in selection, the role of physiology in development, the role of embryology in evolution. Novices in biology often know these mechanisms in isolation from one another. This can be forgiven, as biological instruction often presents material in a way that reinforces the tidy demarcations of contemporary academia. But expert knowledge in biology, and with it, full appreciation of the weight of the evidence for our understanding of the life world, requires seeing how the different aspects of that world, at different levels and at different temporal scales, fit together. The search for biological mechanisms provides a framework, a scaffold, onto and around which the contributions from diverse fields can converge, bringing with them new constraints and perspectives on the space of possible mechanisms.

BIBLIOGRAPHIC DISCUSSION

Many philosophers of science have argued for integrative interfield strategies as opposed to a reductive one. Early work is found in Darden and Maull's (1977) work on interfield theories. William Bechtel (1986) introduced the term *integration* in the very useful introduction and collection of articles that resulted from a conference on integrating scientific disciplines. In that introduction, Bechtel discussed institutional and sociological aspects of scientific disciplines; we neglect that topic here, given our focus on discovering mechanisms. Alison Wylie (2002) discussed the role of multifield perspectives in producing a robust result. Darden (2005), as well as Bechtel and Abrahamsen (e.g., 2007), continue to argue for integrative perspectives in later work. Craver (2007) argues for integration of both mechanism levels and findings from multiple fields in neuroscience.

Philosophers distinguish many kinds of reduction; for an overview, see Brigandt and Love (2008). Here, our topics are not ontological reduction, i.e., what the fundamental furniture of the world is; nor eliminative theory reduction, where one theory is eliminated when another theory explains its domain, see Kemeny and Oppenheim (1956); nor traditional derivational theory reduction, in which one axiomatized theory is deduced from another; see Nagel (1961, ch. 11) and Schaffner (1993). Instead, we concern ourselves with a reductive research strategy that admonishes the scientist always to decompose a system into parts and look down, perhaps down many levels, to find the target mechanism. We argue that looking down is only one of several strategies to employ in mechanism discovery across the biological sciences; it is only a part of an integrative strategy. Wimsatt (1980) also discussed biases in reductive research strategies.

For more on the history of the intralevel, interfield discovery of the protein synthesis mechanism, see Darden and Craver (2002). Hoagland discussed his meeting with Watson in his autobiography (Hoagland 1990, ch. 5). Crick (1996) explicitly credits Linus Pauling with stimulating his interest in weak polar bonds to understand the structure of macromolecules.

For more on the multilevel multifield mechanisms of learning and memory, see Craver and Darden (2001), Craver (2003; 2007). By now, the research program concerning LTP has grown so diverse and multifaceted as to defy comprehensive review. For some internal perspectives on its history and some relatively recent developments, see Bliss, Collingridge and Morris 2004. Schaffner (1993) focuses on parallel work by Eric Kandel on learning and memory in *Aplysia* and emphasizes the reductive aspects of this research program. John Bickle (2003) also argues for a more reductive and less integrative view of this history than recommended here. Jackie Sullivan (2009) discusses the many experimental approaches to the study of LTP and the implications of that multiplicity for the integrative picture sketched here.

Philosophers of biology have debated the nature of the relations between Mendelian genetics and molecular biology for some fifty years. They have proposed a variety of relations between the fields, including reduction (Schaffner 1993), replacement (Hull 1974), and explanatory extension (Kitcher 1984). Darden (2005) discussed and argued against these views and argued for the integrated mechanistic analysis presented here.

The writings on evolutionary theory are extensive. The history of ideas about natural selection can be traced by examining Charles Darwin's views in the *Origin* of 1859, by seeing how Theodosius Dobzhansky integrated findings from Mendelian genetics and cytology with Darwinism and extended it in his seminal book of the evolutionary synthesis, *Genetics and the Origin of Species* (1937), and by examining the reductive view of Richard Dawkins in *The Selfish Gene* (2006). Julian Huxley (1942) named the early twentieth-century achievements in integrating Darwinism and Mendelism in his book *Evolution: The Modern Synthesis.* The standard textbook account, which takes a more phenotypic view than Dawkins' selfish gene view, is Futuyma (2005). Darwin's principle of divergence is found in the *Origin* (1859, pp. 111–25) and discussed by the historian of biology Dov Ospovat (1981).

For more on each of the levels within the mechanism of natural selection: Watson et al. (2007, ch. 9) describes the molecular mechanisms of DNA mutation and repair. Futuyma (2005, ch. 8) discusses the estimation of mutation rates important in evolutionary change. Some definitions of natural selection focus on the downstream effects, rather than on the crucial joint activities between a trait and a critical environmental factor. Such a definition is more general, but it neglects the difference-making stage of the selection mechanism. The standard evolutionary theory textbook, for example, defines natural selection as "*any consistent difference in fitness [reproductive success] among phenotypically different classes of biological entities.*" And further: "The **fitness**—often called the **reproductive success**—of a biological entity is its average per capita rate of increase in numbers" (Futuyma 2005, p. 251; italics and bold in original). This definition of fitness makes fitness more easily measurable, but it is not Darwin's fitness that results from a "struggle for existence" and is captured in the phrase that Darwin adopted in later editions of the *Origin*: "survival of the fittest." Nor does it capture what is important in the organism's activities in a challenging environment, resulting in engineering or (seemingly) designed fitness. For a defense of this view on the importance of a struggle for existence in natural selection by philosophers of biology, see Lennox and Wilson (1994); it is argued for less explicitly in Darden and Cain (1989). For an alternative, more general, view of natural selection that does not require a struggle for existence (and thus isn't a mechanism?), see Lewontin (1970) and Lloyd (2007).

Ernst Mayr's (1942) seminal work on speciation is *Systematics and the Origin of Species*. Jason Byron (previously Baker 2005) discussed isolating mechanisms in speciation. For recent work on speciation genes, see Strain (2011).

The philosopher of science Ben Barros (2008) effectively argues that natural selection is a two-level mechanism; he discusses the Seeley (1986) snail/crab study in more detail. Matthew J. Barker introduced the idea that natural selection is a wide *range* mechanism and also influenced many of the ideas about the mechanism of natural selection presented here. Darden (1986) argues for an early version of the multilevel, integrative view of the evolutionary synthesis that is expanded in this chapter.

Herbert Simon (1996), who shared our interest in understanding reasoning in scientific discovery, discussed the influential idea of nearly decomposable systems. David Rudge (1999; 2001) discusses Kettlewell's peppered-moth case.

11 THE PRAGMATIC VALUE OF KNOWING HOW SOMETHING WORKS

INTRODUCTION: CONTROL

There is a tight connection between mechanistic knowledge and control. When we know how a mechanism works, we know the buttons and levers inside it that might be pushed and pulled to make it work for us. We learn, at least in principle, how we might prevent the mechanism from working, stimulate it into action, and modulate its behavior at will. In some cases, knowledge of biological mechanisms inspires new engineering designs serving novel, practical purposes.

This tight connection between knowing how something works and knowing how to make it work is by now a familiar, and controversial, feature of the biological sciences. Our knowledge of physiological mechanisms has revolutionized medicine by guiding diagnostic reasoning and targeting possible interventions. Such advances are now being replicated at the molecular level, as chemists design chaperones to fold corrupt proteins and chemical agonists and antagonists in order to provoke and prevent myriad chemical mechanisms in cells. Our understanding of genetics has produced an era in which genetically modified organisms are as common on the dining room table as they are in scientific laboratories and in which cloning is a new fact of life. While it would be a gross overstatement to say that all practical advances in biology spring from an understanding of mechanisms (for surely many of our practical advances arise through practical knowhow, serendipity, and trial and error), it is certainly true that most of the practical advances in the domain of biology in the last hundred years—for good or for ill—have been fueled or accelerated by a deepening understanding of life's diverse mechanisms.

In this chapter we focus on a few examples to show how knowing a mechanism helps one devise new ways to control it. Our first example is from ecology, where knowledge of the mechanisms for producing dead zones in the Chesapeake Bay provides a clear basis for designing strategies to alleviate, if not eliminate, a growing environmental catastrophe. Our second example is from genetic medicine, where accumulating knowledge of the mechanisms of cystic fibrosis continues to suggest possible treatment mechanisms, either to cure the disease or to attenuate its deadliest symptoms. Our final example is from neuroscience, where knowledge of genetic, viral, and electrophysiological mechanisms has led to new investigative techniques that can control the activities of brain regions with light. In each case, knowledge of mechanisms guides new strategies for bringing the biological world (and its malfunction or pathology) under our control. Mechanistic knowledge thus brings with it new responsibilities for deciding how that knowledge ought and ought not to be put to use.

ESTUARY MECHANISMS: ELIMINATING DEAD ZONES

Estuaries are enclosed bodies of water at the interface between rivers and streams on the one hand and oceans on the other. These are highly interactive ecosystems and typically stretch over vast watersheds. For example, the Chesapeake Bay watershed, the largest in the United States, covers over 64,000 square miles (103,000 square kilometers). Given their location between the land and the sea, estuaries tend to concentrate nutrients. This concentration process, in which runoff from impervious surfaces, agricultural land, and sewage rapidly enters major waterways and creates nutrient imbalances, is called *eutrophication*. Estuaries around the world face this environmental challenge.

A sustainable estuary ecosystem depends on plant life to provide the necessary oxygen content for other organisms. Fish, shellfish, and all other oxygen-using (aerobic) organisms rely on the combined aeration of plant life and diffusion from winds to keep the waters well aerated. Plants and photosynthetic microbes convert nutrients, sunlight, and water into usable energy and oxygen for aerobic creatures. These producers rely on two main nutrients to grow and survive, nitrogen and phosphorus, usually in the form of nitrates and phosphates. In the mechanism of nutrient flow, the availability of nitrates and phosphates determines the rates of chemical processes that control the levels of dissolved oxygen in the estuary. Because the system continually expends its nutrient resources, a steady flow of usable nutrients must be available to the producers to regulate oxygen levels in the estuary. This is typically achieved through the influx of decaying organics and weathered sediments after precipitation. Runoff collects decaying plant life, animal wastes, and weathered rocks and delivers them into the waterways.

This eutrophication process can lead to a situation in which the oxygen levels in the estuary are chronically low (hypoxia). The runoff changes the nutrient balance of the estuary to encourage algae growth. Ever larger algae blooms choke out subsurface water plants by blocking both sunlight and nutrient flow, depriving the water of one main oxygen source. Then, after the nutrient overload ends and input levels return to normal, the algae population explosion ends and massive die-offs occur. Decomposers then begin their work, and the sheer quantity of dead biomass being decomposed results in a rapid and substantial plummeting of dissolved oxygen. Damage done from this mechanism malfunction leads to expansive dead zones where little to no life can exist. Fish and shellfish kills erupt, which encourages further decomposer oxygen use. Under these circumstances, invasive species thrive, biodiversity decreases, and colonies of toxic bacteria, such as botulism, grow abundantly. This makes the waters even more dangerous for humans and other animals to drink. What was once a thriving estuary ecosystem can quickly become a watery wasteland. In the Chesapeake, expanses of the watershed are, it is believed, now damaged beyond self-regulated repair.

By understanding the mechanisms of eutrophication, scientists and policy advisers are in a position to generate possible means of dealing with the problem of dead zones. By knowing how this mechanism works over time, one is in a position to say which treatments will be palliative (and perhaps ultimately ineffective), and which stand some long-term chance of lessening this environmental crisis.

A number of possible interventions have been proposed to alleviate the dead zones of the Chesapeake. One involves using pumping stations to reoxygenate the water. This treatment falls late in the eutrophication mechanism, as it attempts to counteract the effects of algal blooms. Although this intervention would possibly prevent major fish kills in the short term, it would do nothing to reduce the earlier stages of nutrient influx, or algae and cyanobacteria bloom stages. Eventually, the positive feedback in the eutrophication mechanism would overwhelm these palliative efforts. Such a proposal treats one of the symptoms without eliminating the underlying disease, and as a result without eliminating the other symptoms.

Another suggested treatment method involves reducing algae through predation. Introducing predatory daphnids during algae blooms could prevent blooms from reaching the scale that would adversely affect water quality. This treatment intervenes at an earlier stage in the mechanism than oxygen replacement, but it still has drawbacks. Because it does not address the increased nutrient loads, a growing population of daphnids would be required to control algal growth. Such an intervention runs the risk of allowing daphnid populations to grow out of control; they might, for example, overgraze the algae even outside of the bloom times and reduce the algal population too dramatically. Such an intervention also runs the inherent risk of adding new creatures in abundance to a food chain.

Given how this mechanism works, it is clearly preferable to treat key earlier stages, those with a greater number of downstream consequences, than to try merely to counteract the symptoms of this more fundamental problem. The most effective intervention would be to change the nutrient loads themselves. One might try to remove the offending nutrients, but there has thus far been no successful method of removing such a large quantity of nutrients after they have entered the waterway. The most accessible and effective intervention, it would seem, is at the earliest stage of the eutrophication mechanism. Such an intervention would slow or stop the nutrient loads before they enter the estuary. In the case of the Chesapeake, such an intervention would work to prevent runoff from factory chicken farms and from the use of fertilizer in suburban lawns.

Of course, knowledge of the mechanisms of eutrophication does not tell us how best to bring about the policy changes required for the best solution to this environmental problem. However, perhaps if scientists and policymakers become more adept at communicating their knowledge of the underlying mechanisms, they will become all the more convincing in their efforts to convince the required individuals and institutions to make the changes required to address most effectively our deepest environmental conundrums.

Why should environmental scientists follow such a mechanistic approach? A reasonable answer is that a mechanistic outlook has led to similar advances in the medical sciences. We turn now to just such an example.

CYSTIC FIBROSIS MECHANISMS AND DESIGN OF THERAPIES

Cystic fibrosis is an inherited disease that affects the lungs and the digestive systems of about 70,000 people worldwide, many of whom are children. As recently as the 1950s, children with cystic fibrosis could not reasonably expect to live long enough to attend school. Now, people with the disease regularly live well beyond their third decade. Many of these advances depended fundamentally on knowledge of the mechanisms by which the disease progresses.

Clinicians first identified cystic fibrosis as a disease in the early twentieth century when they realized that a number of unhealthy people shared a cluster of correlated symptoms in the lungs, the pancreas, and the sweat glands. They had recurrent respiratory infections, elevated levels of chloride in sweat, and insufficient pancreatic enzymes. These symptoms were ultimately connected to problems in the epithelial tissues of those organs and glands. Population genetic studies of families with CF indicated that the disease is hereditary, not sex-linked, and requires two copies of the mutant gene to produce the disease symptoms; carriers with one copy are not sick. (In more technical genetic terms, it is an autosomal recessive disease.) The fact that patients with the disease tend to have elevated levels of salt in their sweat led researchers to surmise that perhaps the responsible gene is typically associated with the transport of chloride ions into and out of cells. However, the specific nature of the defect in epithelial tissues was unknown until the 1980s.

Researchers used new molecular biological techniques developed in the 1970s and 1980s to identify the defective gene. In 1989 its discoverers gave it an unwieldy name—cystic fibrosis transmembrane conductance regulator, or CFTR for short. The gene is large, with approximately 180,000 base pairs on the long arm of chromosome 7. It produces a large protein with 1480 amino acids, organized into several different functional domains. The CFTR protein produced by the CFTR gene spans the cell membrane in epithelial tissues. It controls the concentration of chloride ions in the cell. This function explains the salt (NaCl) imbalances in the tissues with a defective CFTR gene. Researchers found six different kinds of mutations in the gene that result in failure at different stages of the protein synthesis mechanism and produce variants of the disease.

In the early 1990s it looked as if it would be easy to conquer cystic fibrosis. The gene had been discovered. The relevant mechanisms and mutations would be understood; gene therapies would replace the defective gene, at least in the lungs, and prevent the worst of the symptoms. The defective gene is the first stage of the disease mechanism and thus seemed to be the most promising target for therapy. However, several problems got in the way. The gene is large and does not easily fit into the vectors available for delivering the gene to the (even relatively accessible) lung cells. As with any gene therapy, the inserted gene must arrive in a safe location so as not to disrupt other mechanisms. A sufficient number of cells have to incorporate the gene so that enough functional protein can be produced. Also cellular regulatory signals need to properly turn on the gene but not overproduce the protein. Although research continues, these problems for gene therapy remain.

So, consider the next stage of the mechanism, the one after the gene itself, as a target: the messenger RNA. The CFTR gene contains not only the coding sequences that eventually direct the ordering of amino acids during protein synthesis, but also spacer segments, called introns. A cell organelle called a spliceosome processes the pre-mRNA; it snips out the introns and binds together the remaining coding segments (exons) to produce the messenger RNA. Researchers have succeeded in inserting a minigene into the DNA of human lung tissue grafted onto a mouse. The minigene has the correct coding segment rather than the common three base mutation. The minigene is expressed at the same time as the CFTR gene, thereby overcoming one of the barriers to gene therapy. Then the splicing machinery is induced to put the correct segment into the processed messenger RNA rather than the mutant segment. Some success in the mouse system makes this look promising. However, it is still a long way from human clinical trials.

Currently, a primary area for targeted drug therapies for CF is the next stage of the mechanism, protein synthesis. Although some of the mutations result in no protein synthesis, the mutation in about 90 percent of patients with cystic fibrosis in the US produces a malformed protein. The mutation is just a three-base deletion in the DNA, and it results in just one missing amino acid: phenylalanine at position 508 (Delta F 508). This absence results in a misfolded protein that does not get implanted into the cell membrane. Normally, the cellular machinery degrades misfolded proteins, but not all of this mutant protein is degraded.

Current efforts in drug therapy are directed to finding drugs that aid in rescuing the undegraded, misfolded protein by refolding it. The protein then inserts itself into the cell membrane and restores, to some extent, the transport of chloride ions. A robotic process has screened millions of compounds for their effects on the misfolded protein, and researchers have found some promising drug candidates. In contrast to this random screening approach, rational drug therapy uses mechanistic knowledge to guide treatment discovery. Researchers use a more detailed understanding of so-called chaperone molecules that help the CFTR protein to fold properly.

The discovery of the roles played by additional molecules that interact with the CFTR protein helps to explain the role that additional genes, called modifier genes, may play to explain a puzzling phenomenon about the relation between genotype and phenotype. That puzzling phenomenon is this: patients with the same two mutations can still vary in the severity of symptoms of the disease. One hypothesis is that this difference is due to different modifier genes that normally play a role in protein folding. If appropriate chaperones can be inserted or induced to form, then the misfolded protein might be restored to functionality.

Thus we see that each stage of the mechanism from the defective gene to the defective protein is a candidate target for therapy. In addition, one can ask: Is there some other mechanism, besides repairing the broken mechanism, that can restore chloride transport function? This avenue is being explored currently. One can also simply focus on palliative mechanisms to help to alleviate the symptoms without eliminating the root cause of the symptoms. More specifically, one can seek drugs that help to prevent the lung infections from which cystic fibrosis patients typically die.

Many of the details about the mechanisms by which a defective CFTR protein produces the symptoms of CF are unknown. It is known that, as cystic fibrosis patients age, they become more susceptible to particular strains of bacteria that are more resistant to treatment. Various hypotheses as to how to fill these black boxes abound. As a recent review article said: "So far, a unifying mechanism responsible for the vast clinical expression of the disease in the CF airway has not been identified" (Chmiel and Davis 2003, p. 173).

There are many competing hypotheses to explain why the mucus in the airways of cystic fibrosis patients becomes so thick and so susceptible to bacterial infections. Several of these posit that malfunctioning chloride transport leads to salt imbalances of salt homeostasis or abnormal water absorption, which in turn produces thicker mucus. However, new evidence points to a possible malfunction in the immune response. Neutrophils, a type of white blood cell, are recruited to fight bacteria. CFTR defective cells also have defective neutrophil regulation, and perhaps this leads to an overabundance of neutrophils. This mechanism is not well understood, although some of the entities and their activities have been identified. Roughly, the remnants of broken-down neutrophils, especially their DNA, produce thick mucus that becomes a site for bacteria to colonize. So, if the physician wants to prevent the bacterial infections, she might target her therapeutic efforts at neutrophil regulation mechanisms. These mechanisms appear much later in the disease process, and they provide different targets than the CFTR protein biosynthesis mechanism. A complementary, higher-level perspective is to use postural and percussive methods to shake the excess mucus and void it from the lungs by coughing.

The hypothesis that CF results from an overactive immune response suggested to some that perhaps anti-inflammatory drugs would benefit CF patients. If this hypothesis were not true, one would expect anti-inflammatory drugs, such as ibuprofen, to be harmful to lungs susceptible to infections. The normal inflammatory response, which recruits neutrophils to the site of an infection, would help to fight the bacteria. One would not want to suppress it with ibuprofen. However, if this overactive immune response hypothesis is correct, ibuprofen would tend to lessen the hyperactive immune response and make infection less likely. This treatment has been successful in clinical trials. Thus, a different mechanism, coming later in the progression of the disease, is a possible target for controlling one disease symptom.

At a still higher-level perspective, many people with CF become weak and thin because they fail to maintain an adequate body weight. Such consequences of the disease are thought to be downstream from its primary mechanism of action, but people with CF frequently benefit from nutritional interventions with high calorie diets and, in younger or more extreme cases, tube feeding.

In this case, many different mechanisms provide candidate targets for therapy. Knowledge of the gene defects, of the different ways the protein synthesis mechanism malfunctions, and of the secondary mechanisms during the progression of the disease—all suggest possible therapies, operating at levels from molecules to whole organisms. One might have thought that, for a disease due primarily to a single defective gene, a simple genetic fix would quickly be available. Sadly that's not how things have turned out so far. As a result, researchers have had to rely on their knowledge of the various stages of the normal mechanisms at multiple levels, the specific ways in which they break, and the different mechanisms that come into play as the disease progresses. All these open potential doors to the discovery of new drugs and new therapies.

OPTOGENETICS: CONTROLLING THE BRAIN WITH LIGHT

Our last example of translational progress through mechanistic knowledge is the recent development of optogenetic techniques. Although this example of progress in neuroscience is pregnant with possible therapeutic applications in the future, its primary import in present-day neuroscience is as an advance in experimental control over neural systems.

Optogenetics is a technique for manipulating the chemical and electrical properties of biological mechanisms with light. In neuroscience specifically, the technique is used to manipulate the electrophysiological properties of neurons. The technique trades on the ability of researchers to transplant genes for light sensitive ion channels from bacteria into the cells of another organism (such as a mouse, a monkey, or, someday, a human). The researchers first prepare a gene construct with regulatory units (promoters and inhibitors) that will cause the gene to be expressed only in cells with a specific chemical or genetic profile, with particular morphologies, or with specific forms of connectivity. They then use a virus to deliver the construct to a particular location in the brain. In the cells with the right combination of promoters and inhibitors, protein synthesis and delivery mechanisms assemble the channels and insert them into the cell membrane. Researchers then insert a fiber optic cable into the brain near the target region. They deliver light of a specific wavelength through the cable, activating the newly inserted channels. The channels then change their conformation to an open state, allowing ions to flow across the membrane. This ionic current can be used to raise or lower the membrane potential of the cell, and so to produce or retard electrophysiological signals. Optogenetics, in a nutshell, places the cells' electrophysiological properties under the direct control of the experimenter.

Optogenetics is typically introduced as an advance over two traditional intervention techniques: electrophysiological interventions and pharmacological interventions. In a traditional electrophysiological intervention, one uses an electrode to deliver current to a cell or a brain region. In the laboratory, researchers can intervene on single cells and deliver currents that are comparable to those that typically drive a cell. For populations of cells in a living organism, however, the technique is quite a bit less surgical: a current is passed from the electrode into the brain tissue. The consequence is a large and simultaneous shift in the activity level of all of the cells and other tissues within a given volume around the tip of the electrode. Pharmacological interventions in this kind of experiment typically involve the exposure of a brain region to a chemical agonist or antagonist that either stimulates or inhibits cells of a given type. Such interventions can give the researcher a good deal of control over the types of cells they target (sorted on the basis of neurotransmitter and receptor types used by the target cells). Pharmacological interventions in large-scale neural systems (such as the brain of an awake and behaving organism) sometimes lack system specificity (they activate many brain systems at once if administered via the bloodstream, for example) and in many cases lack the temporal precision optogenetics makes available. Optogenetics thus affords, in many cases, interventions that target very specific kinds of cells with the temporal precision required to mimic neuronal firing patterns at will. The technique has already been used to explore the mechanisms of anxiety, classical conditioning, cocaine addiction, depression, locomotion, sleep and waking, and sensory processing in non-human organisms. Given its many virtues, one can expect the technique to expand its domain of application considerably in coming years.

What matters for present purposes, however, is not what the technique allows researchers to do, but rather how mechanistic knowledge in biology made this experimental advance possible. Consider the diverse kinds of mechanisms one must understand in order to conceive and build reliable optogenetic methods. One must first know that bacteria have light-sensitive channels, and one must know that the channels work by trafficking ions across the cell membrane. One must know that these channels are synthesized by protein-synthesis mechanisms; one has to know the location of the relevant gene; one has to know how gene regulatory elements work; and one has to be able to match the gene regulatory elements to the specific cellular environments of the target cell. One must understand viruses well enough to use them to inject the relevant genetic material. And of course, one must understand the electrophysiological mechanisms controlling the activities of neurons well enough to understand that one can control neurons by controlling the flux of ions across the membrane. In short, one could not begin to plausibly dream of a technique such as optogenetics unless one already knew quite a bit about mechanisms involving genes, viruses, and neurons. Indeed, one might argue that the capacity to use our mechanistic knowledge to produce novel effects in the world is the most convincing evidence one might have for the truth of one's schemas.

CONCLUSION

Francis Bacon defined science as the search for causes and the production of effects. In this chapter we emphasize the relationship between these two aims. When one knows how a mechanism works, one can know how to work with mechanisms to produce new effects. This is true at many levels of biological research. Knowledge of ecological mechanisms, such as eutrophication, allows one to devise strategies to slow or prevent an environmental catastrophe (if only the political will can be mobilized). Knowledge of disease mechanisms, such as those underlying the symptoms of CF, allows researchers to reason their way to new potential therapies. And knowledge of biological mechanisms provides bioengineers with raw materials to fashion new instruments, such as optogenetics, that afford greater experimental control and the production of effects that are otherwise unimaginable. Perhaps not all knowledge is power, but mechanistic knowledge certainly is. And with this power comes new responsibilities: for the once intractable problems we solve, for the new problems we create, and for the now solvable problems we fail to address.

BIBLIOGRAPHIC DISCUSSION

The eutrophication case is taken from unpublished paper by Renard J. Sexton, presented at the 2007 meeting of the International Society for History, Philosophy, and Social Studies of Biology in Exeter, UK. Much information about dead zones in Chesapeake Bay is available from the publications of the Chesapeake Bay Foundation (http://www.cbf.org).

The cystic fibrosis case is discussed in more detail in Darden (forthcoming). A useful collection of articles on the disease is Kirk and Dawson (2003); the discovery of the CFTR gene is recounted in Drumm (2001). Wang et al. (2006) document the new work on chaperones and CFTR protein folding. Historian of science Susan Lindee is working on a book about the failures of gene therapy for cystic fibrosis; she presented an excerpt, "A Disease About to Disappear: Gene Therapy for CF," in a session on "Genes and Mechanisms in the Case of Cystic Fibrosis: Philosophical, Historical and Social Perspectives," at the History of Science Society meeting in Montreal, 2010. We thank her for help with ideas about and sources on CF. For more on the topic, see Lindee and Mueller (2011).

Boyden et al. (2005) is the first paper reporting the use of optogenetic methods. For some more recent reviews, including discussion of the stunning application of this technique, see Deisseroth (2010; 2011).

12 CONCLUSION

DISCOVERING MECHANISMS

There can be no single, monolithic philosophy of discovery. There is no non-trivial insight that applies equally to the discovery of a new star, of a novel proof, of an intracellular signaling cascade, of a new method for staining neurons, and of a new way to test a hypothesis. Different questions and strategies are useful for discovering different kinds of things. The product, as we have remarked throughout the book, shapes the process of discovery. A plausible first step toward a richer understanding of the nature of scientific discovery, then, is to restrict the domain to a specific kind of discovery. Our focus is on the discovery of biological mechanisms.

Mechanisms are our focus because much of biology and its history involve constructing, evaluating, and revising our understanding of mechanisms. Physiologists study mechanisms in the body at large and small scales, from the circulation of the blood to the filtration of water in microscopic tubules of the kidney. Evolutionary biologists study the mechanism of natural selection and the isolating mechanisms of speciation. Ecologists study mechanisms that maintain and disrupt ecosystems such as the Chesapeake Bay. Geneticists study the mechanisms underlying patterns of inheritance. Neuroscientists investigate the mechanisms of spatial memory, the propagation of action potentials, and the opening and closing of ion channels in neuronal membranes. Molecular biologists study the mechanism of protein synthesis and myriad mechanisms of gene regulation. Medical scientists investigate the genetic and physiological mechanisms that give rise to cystic fibrosis. We focus on the processes by which scientists discover mechanisms because the bulk of biology is a search for mechanisms in the living world.

As we have stressed, three primary motivations explain why fascination with mechanisms occupies such a central place in the biologist's self-conception. First, knowing the mechanism usually allows one to predict how the phenomenon will behave. If one knows how something regularly works, one can often say how the mechanism would work if it were placed in similar conditions. Second, and related, knowing the mechanism for a phenomenon makes the phenomenon intelligible. In some cases, one can literally see how the mechanism works from beginning to end. In other cases, one relies upon other modes of intelligibility, such as understanding mathematical relationships among variables describing system components. Finally, knowing the mechanism potentially allows one to intervene into the mechanism to produce, eliminate, or change the phenomenon of interest. Biological mechanisms, in other words, fascinate us in part because we are interested in controlling our world.

It is not surprising then to find that biologists have devised methods and strategies for discovering mechanisms. Of course, scientists are rarely in the business of making the abstract structure of their reasoning explicit. And sometimes, we must admit, the relevant bit of reasoning comes into view only with time, as we look back at a discovery episode with some historical distance and with the benefit of scientific hindsight. Yet the history of science, viewed from this perspective, contains within it patterns of reasoning, strategies for describing and discovering mechanisms, that implicitly shape everything biologists do: the choice of model organisms, the selection of experimental apparatus, the design of experiments, and the interpretation of results. One cannot write the history of modern biology without recognizing how thoroughly this mechanistic perspective shapes what biologists do, how they think about their subject matter, and the norms they enforce on one another in the search for mechanisms.

Our first goal, then, has been to provide a descriptive account of progress in mechanistic research programs: to show how mechanistic research programs change over time, to discuss the evidential constraints that drive those changes, and to show how background knowledge shapes the available space of possibilities. Our perspective on the search for mechanisms involves: constructing a space of how-possibly schemas; sorting through the candidates to find the most plausible schema; gathering evidence through observation and experiment to shape the space of possible mechanisms; diagnosing how one's hypothesized schema fails; and mining the character of that anomaly for clues about how to revise the schema or where to look in the space of possible mechanisms for a suitable replacement. The result is a descriptive picture of the tools and reasoning strategies by which mechanistic research programs make progress.

But we also have an instructive agenda. The diverse discoveries we consider from the history of biology offer some insight into the questions, constraints, and strategies one might fruitfully consider in trying to discover a mechanism. These historical exemplars are offered not as empirical evidence for the general utility of these strategies; no single case could support such general claims. Nor are they offered as apparent insight into the inner workings of the scientific mind; we allow that such facts are often opaque to us. Rather these historical cases are offered as individual demonstrations of how a general strategy for reasoning about mechanisms can be applied to solve a particular kind of problem that arises during a discovery episode. These strategies do not guarantee success; they can be offered only as moves one might make, things one might consider, heuristics one might deploy in trying to specify and solve a mechanism discovery problem. We offer a tool kit, not a road map.

In our view the search for mechanisms is inherently a pluralistic endeavor, requiring one to integrate findings from diverse fields within the biological sciences. Because different fields specialize in different constraints, work at different levels, investigate different time scales, and, most generally, come at the world from different perspectives, they are each potentially in the position to contribute uniquely to filling out a mechanism schema. Given that biological mechanisms are made of diverse parts, with myriad forms of organization, spanning multiple levels of mechanisms within mechanisms, biology cannot afford to privilege the discoveries of researchers in one or a few subspecialties. Mechanistic biology demands interfield integration.

As we noted above, our approach to the topic of discovery begins by narrowing the task. We focus on the discovery of mechanisms rather than other kinds of discovery. We also focus on biology rather than other sciences. It is an open question to what extent the perspective on mechanistic discovery we offer here will be useful in other sciences. The search for mechanisms is clearly not restricted to the biological world. There are no doubt mechanisms in the sense discussed in this book in sciences as diverse as geology, chemistry, meteorology, and many areas of physics. Talk of mechanisms can also be found in more contentious domains. Psychologists routinely appeal to cognitive mechanisms: attentional mechanisms, perceptual mechanisms, memory mechanisms, and the like. Many economists and social scientists talk freely about the mechanisms of decision-making, market mechanisms, and mechanisms of social change. Perhaps there are areas of science where nonmechanistic patterns of explanation are accepted, where distinct norms of explanation and discovery apply. Those areas of science would clearly require a different account of discovery altogether. We leave it to others more familiar with these sciences to assess whether what we say about mechanism discovery in biology is applicable in other sciences and whether the notion of mechanism developed here plausibly expresses the sense of the term as it is deployed in those sciences where it does, in fact, occupy a similarly central place.

We also leave it as an open question whether there are legitimate, nonmechanistic forms of explanation in biology. If there are limits to mechanistic explanation, and if these limits can be discovered empirically, then one reasonable way to move forward is as follows: continue to search for mechanisms. When the search fails recalcitrantly, then consider nonmechanical forms of explanation. Perhaps in the meantime, the opponent of this mechanistic perspective can take some solace in the fact that the search for mechanisms is at least in many contexts well-suited for delivering the primary aims of science: prediction, explanation, and control.

SUMMARIES

Our discussion builds from our broad characterization of mechanisms in Chapter 2. This characterization should be read not so much as a definition but as a most abstract schema of a mechanism, as the sort of thing that might be filled with specific entities, activities, and organizational features to make up any of the myriad mechanisms found in biology textbooks. The placeholders in this characterization serve as something of a checklist to evaluate the completeness of one's description of a mechanism, to evaluate whether it includes all of the entities, activities, setup and termination conditions, whether it describes the active relations between successive stages that constitute the mechanism's productive continuity, whether it describes the relevant features of the mechanism's spatial and temporal organization, and whether it describes the mechanism at all the relevant levels for the purposes at hand.

Mechanism schemas are the explicit focus of Chapter 3. We distinguish mechanism schemas, which can be expanded as necessary to reveal all the relevant features of a mechanism for a given purpose, from mechanism sketches, which have black boxes (that is, gaps in the mechanism's productive continuity) or gray boxes (functional roles for which the underlying mechanism is not known). A schema can have the status of merely a how-possibly description of a mechanism or it can have the status of being how-actually-enough for the purposes at hand. Schemas can also vary considerably in their scope, applying to all organisms, or perhaps only to a handful of cells in a single species. Such schemas can be represented in any number of ways, be it with movies, with pictures, with words, and with equations; but what makes it a mechanism schema is that it purports to describe a mechanism that produces, underlies, or maintains a phenomenon of interest.

As Chapter 4 makes clear, a puzzling phenomenon is often the beginning point in the search for a mechanism. How one chooses to characterize the phenomenon places strong constraints on what one will accept as an acceptable mechanistic explanation for it. A complete characterization of the phenomenon will include: its setup and start conditions (if such an initiating trigger exists), as well as its termination point; conditions that inhibit it; conditions that modulate its behavior; any nonstandard conditions in which it may occur or change; and any by-products that accompany its occurrence. These all provide clues in the search for its mechanism. Sometimes, however, one is forced to revise one's description of the phenomenon as the search for its mechanism proceeds. Most dreadfully one might learn that the phenomenon does not in fact exist. Or one might learn that the phenomenon one set out to study is, in fact, two distinct phenomena (cases of splitting), or that it is only part of a much more inclusive phenomenon (lumping). More common, however, is the finding that one's understanding of the phenomenon has to morph to fit the details one discovers about the mechanism that produces, underlies, or maintains it.

Characterizing the phenomenon begins the iterative stages in the search for a mechanism. Next are ongoing and piecemeal efforts of constructing, evaluating, and revising mechanism sketches and schemas. One can imagine moving in a path through a space to fill in the black and gray boxes of sketches and to find how possibly, then how plausibly, and then finally how actually the mechanism works as constraints on the space accumulate. Numerous historical episodes of such successful (and some unsuccessful) discovery paths supply questions, constraints, and strategies that are potentially useful in traversing such a path.

Chapter 5 suggests strategies for schema construction to help formulate the space of possible and plausible schemas for mechanisms that produce, underlie, or maintain the phenomenon of interest. To obtain the overall organization of the mechanism, one might retrieve and map an analogue or invoke an abstract mechanism type. Table 5.1 has a preliminary list of some types of mechanisms invoked in the history of science. When no prior overall organization is available, a researcher may be able to construct a mechanism schema by putting together modules or forward/backward chaining. If one doesn't even know enough to use any of these schema construction strategies, then a fallback position is to go back to poking and prodding the phenomenon, with the hope of localizing it sufficiently to begin investigating its mechanism.

With a how-possibly schema or sketch in hand, one begins to evaluate it. Chapters 6, 7, and 8 explored different aspects of mechanistic schema evaluation. In Chapter 6 we discussed the various virtues that one hopes to maximize in constructing a mechanism schema. In addition to traditional theoretical virtues (empirical adequacy, fertility, generality, etc.), we explored in depth some vices specific to the search for mechanisms, namely, superficiality, incompleteness, and incorrectness.

Chapter 7 utilizes the extended stages of William Harvey's discovery of the circulation of the blood to extract specific constraints on the space of possible mechanisms. Mechanism components and stages must satisfy the constraints of having the proper locations, structures, abilities, activities, timing, productivity, continuity, and roles, as well as the proper fit into the global organization of the mechanism. Harvey's demonstration that the blood circulates uses all of these strategies and provides an exemplar of mechanism evaluation.

Chapter 8 steps back from the detailed search for mechanism components to see how to use experiments to evaluate cause-and-effect relations; later the mechanistic details fill the cause-effect sketch. Experiments on mechanisms require an experimental system running the mechanism. The researcher intervenes at one point and detects the effect at another. A successful result reveals a cause-and-effect relation among some parts of the mechanism. Careful control of conditions yields a clean result. Types of interlevel experiments include inhibition, stimulation, and activation. Additional experiments for testing mechanisms include by-what-activity experiments, by-what-entity experiments, the use of a series of experiments for testing organization, and the need to prepare the experimental system to reveal specific activities of the mechanism.

Again, the discovery of a mechanism is a piecemeal iterative process, not a linear march from constructing a schema to demonstrating its adequacy. Often anomalies turn up, and some require revision of a hypothesized mechanism schema. Chapter 9 analyzes responses to types of anomalies. Here's a list, ordered from less to more severe challenges to a hypothesized schema: experimental error, data analysis error, monster anomaly, special case anomaly, model anomaly, and falsifying anomaly. The mechanistic perspective provides useful resources in resolving monster, special case, and model anomalies.

Chapter 10 lays out a broader perspective on the role played by scaffolding mechanisms to integrate findings from different fields and levels, as well as across longer time scales. Results from two or more fields sometimes contribute to the discovery of a mechanism at a single level. In other cases researchers strive for interlevel integration, showing how mechanisms within mechanisms underlie a phenomenon. Temporally extended mechanisms integrate numerous submechanisms operating sequentially or simultaneously to produce a phenomenon. Contextual constraints from these wider contexts supply coherence checks on the adequacy of proposed subcomponents in the larger scaffolds.

Once a scientist discovers a mechanism, it becomes available to achieve goals of science: supplying understanding, giving explanations, making predictions, and providing the potential to control outcomes of interest to us. In Chapter 11, examples from medicine, environmental policy, and bioengineering illustrate the broad reach in the kind of control afforded by understanding the appropriate mechanism.

Science, as we have suggested repeatedly, is an engine for making discoveries. Mechanistic science is fueled by empirical findings about the domain in which the mechanism works and by the questions and constraints that scientists recognize as guides in forming adequate mechanism schemas. In this book, we have attempted to make some of these questions, constraints, and strategies explicit as a stimulus to clear thinking about the aims of biological research and about how different parts of biology are assembled into an integrated picture of life. Indeed, if we are right, nothing in biology makes any sense without the idea that biologists are searching for mechanisms.

REFERENCES

Allchin, Douglas. 1997. "A Twentieth-Century Phlogiston: Constructing Error and Differentiating Domains." *Perspectives on Science* 5:81–127.

———. 1999. "Negative Results as Positive Knowledge, and Zeroing in on Significant Problems." *Marine Ecology Progress Series* 191:301–05.

———. 2002. "Error Types," *Perspectives on Science* 9:38–58.

Allen, Colin, Marc Bekoff, and George Lauder, eds. 1998. *Nature's Purposes: Analyses of Function and Design in Biology*. Cambridge, MA: MIT Press.

Allen, Garland E. 1978. *Thomas Hunt Morgan*. Princeton, NJ: Princeton University Press.

———. 2005. "Mechanism, Vitalism and Organicism in Late Nineteenth and Twentieth Century Biology: The Importance of Historical Context." In *Studies in History and Philosophy of Biological and Biomedical Sciences*. Carl F. Craver and Lindley Darden, eds. Special Issue: "Mechanisms in Biology" 36:261–83.

Anand, B. K. and J. R. Brobeck. 1951. "Localization of a 'Feeding Center' in the Hypothalamus of the Rat." *Proceedings of the Society for Experimental Biology and Medicine* 77:323–25.

Andersen, Holly. 2011a. "The Case for Regularity in Mechanistic Causal Explanation." *Synthese*, doi:10.1007/s11229–011–9965-x.

———. 2011b. "Mechanisms, Laws, and Regularities." *Philosophy of Science* 78 (2): 325–31.

Aristotle. 1912. *De Partibus Animalium*. Translated by William Ogle. In *The Works of Aristotle*, vol. 5. J. A. Smith and W. D. Ross, eds. London: Oxford University Press, Clarendon Press.

Avery, Oswald T., C. M. MacLeod, and Maclyn McCarty. 1944. "Studies on the Chemical Nature of the Substance Inducing Transformation of Pneumococcal Types." *Journal of Experimental Medicine* 79:137–58. Reprinted In *Classic Papers in Genetics*. J. A. Peters, ed. 1959. pp. 173–92. Englewood Cliffs, NJ: Prentice Hall.

Axelrod, Julius and Georg Hertting. 1961. "The Fate of Titrated Noradrenalin at the Sympathetic Nerve Ending." *Nature* 192:172–73.

Axelrod, Julius and Robert Tomchick. 1961. "Enzymatic O-methylation of Epinephrine and Other Catechols." *Journal of Biological Chemistry* 233:702–05.

Axelrod, Julius, Georg Hertting, and L. Gordon Whitby. 1961. "Effect of Drugs on the Uptake and Metabolism of 3H-norepinephrine." *Journal of Pharmacology and Experimental Therapeutics* 134:146–53.

Axelrod, Julius, Georg Hertting, Irwin Kopin, and L. G. Whitby. 1961. "Lack of Uptake of Catecholamines after Chronic Denervation of Sympathetic Nerves." *Nature* 189: 66–68.

Bacon, Francis. [1620] 1960. *The New Organon*. New York: Bobbs-Merrill.

———. [1626] 2008. *The New Atlantis*. In *Francis Bacon, The Major Works*. Brian Vickers, ed. New York: Oxford University Press.

Baker [now Byron], Jason M. 2005. "Adaptive Speciation: The Role of Natural Selection in Mechanisms of Geographic and Non-geographic Speciation." In *Studies in History and Philosophy of Biological and Biomedical Sciences*. Carl F. Craver and Lindley Darden, eds. Special Issue: "Mechanisms in Biology" 36:303–26.

Baltimore, David. 1970. "Viral RNA-dependent DNA Polymerase." *Nature* 226:1209–11.

Barros, D. Benjamin. 2008. "Natural Selection as a Mechanism." *Philosophy of Science* 75:306–22.

Beatty, John. 1995. "The Evolutionary Contingency Thesis." In *Concepts, Theories, and Rationality in the Biological Sciences*. James G. Lennox and Gereon Wolters, eds. pp. 45–81.

Bechtel, William, ed. 1986. *Integrating Scientific Disciplines*. Dordrecht: Nijhoff.

———. 2006. *Discovering Cell Mechanisms: The Creation of Modern Cell Biology*. Cambridge Studies in Philosophy and Biology. New York: Cambridge University Press.

———. 2008. *Mental Mechanisms: Philosophical Perspectives on Cognitive Neuroscience*. New York: Routledge.

Bechtel, William and Adele Abrahamsen. 2005. "Explanation: A Mechanist Alternative." In *Studies in History and Philosophy of Biological and Biomedical Sciences*. Carl F. Craver and Lindley Darden, eds. Special Issue: "Mechanisms in Biology" 36:421–41.

———. 2007. "In Search of Mitochondrial Mechanisms: Interfield Excursions between Cell Biology and Biochemistry." *Journal of the History of Biology* 40:1–33.

———. 2010. "Dynamic Mechanistic Explanation: Computational Modeling of Circadian Rhythms as an Exemplar for Cognitive Science." *Studies in History and Philosophy of Science* 41:321–33.

Bechtel, William and Robert C. Richardson. 1993. *Discovering Complexity: Decomposition and Localization as Strategies in Scientific Research*. Princeton: Princeton University Press. 2nd ed. 2010. Cambridge, MA: MIT Press.

Bender, Edward A., Ted J. Case, and Michael E. Gilpin. 1984. "Perturbation Experiments in Community Ecology: Theory and Practice." *Ecology* 65 (1): 1–13.

Benfey, O. T. 1958. "August Kekulé and the Birth of the Structural Theory of Organic Chemistry in 1858." *Journal of Chemical Education* 35:21–23.

Bernard, Claude. 1850. "Action Physiologique des Venins (Curare)." *Comptes Rendus des Séances et Mémoires de la Société de Biologie* t.1 1849 (1850):90.

———. 1851a. "Action du Curare et de la Nicotine sur le Système Nerveux et sur le Système Musculaire." *Comptes Rendus des Séances et Mémoires de la Société de Biologie* t.2 1850 (1851):195.

———. 1851b. "New Experiments on the Woorara Poison." *Lancet* 57 (1437): 298–300. (Originally published in the *Lancet* as vol. 1, issue 1437, March 15, 1851.)

———. 1864. "Études Physiologiques sur quelques Poisons Américaines. I. le Curare." *Revue des Deux Mondes* t.53:164–90.

———. [1865] 1957. *An Introduction to the Study of Experimental Medicine*. New York: Dover.

Bernard, Claude with J. Pelouze. 1850. "Recherchés sur le Curare." *Compte Rendu des Séances de L'Académie des Sciences* t.31:533–37.

Bickle, John. 2003. *Philosophy and Neuroscience: A Ruthlessly Reductive Approach*. Dordrecht: Kluwer Academic Publishers.

Birmingham, A. T. 1999. "Waterton and Wouralia." *British Journal of Pharmacology* 126 (8): 1685–90.

Black, John. 1999. "Claude Bernard on the Action of Curare." *British Medical Journal* 319 (7210): 622.

Bliss, T. V. P., G. L. Collingridge, and R. G. M. Morris. 2004. *Long-term Potentiation: Enhancing Neuroscience for 30 Years.* Oxford: Oxford University Press.

Bliss, Tim V. P. and A. R. Gardner-Medwin. 1973. "Long-lasting Potentiation of Synaptic Transmission in the Dentate Area of the Unanaesthetized Rabbit Following Stimulation of the Perforant Path." *Journal of Physiology* 232:357–74.

Bliss, Tim V. P. and T. Lømo. 1973. "Long-lasting Potentiation of Synaptic Transmission in the Dentate Area of the Unanaesthetized Rabbit Following Stimulation of the Perforant Path." *Journal of Physiology* 232:331–56.

Bliss, Tim V. P., A. R. Gardner-Medwin, and T. Lømo. 1973. "Synaptic Plasticity in the Hippocampal Formation." In *Macromolecules and Behavior.* G. B. Ansell and P. B. Bradley, eds. pp. 193–203. London: Macmillan.

Boas, Marie. 1952. "The Establishment of the Mechanical Philosophy." *Osiris* 10:412–541.

Bogen, James. 2005. "Regularities and Causality; Generalizations and Causal Explanations." In *Studies in History and Philosophy of Biological and Biomedical Sciences.* Carl F. Craver and Lindley Darden, eds. Special Issue: "Mechanisms in Biology" 36:397–420.

———. 2008a. "Causally Productive Activities." *Studies in History and Philosophy of Science* 39:112–23.

———. 2008b. "The Hodgkin–Huxley Equations and the Concrete Model: Comments on Craver, Schaffner, and Weber." *Philosophy of Science* 75:1034–46.

———. 2011. "'Saving the Phenomena' and Saving the Phenomena." *Synthese* 182 (1): 7–22.

Bogen, James and James Woodward. 1988. "Saving the Phenomena." *Philosophical Review* 97:303–52.

Bois-Reymond, Emil du. 1848. *Untersuchungen über Thierische Eelektricität.* Berlin: G. Reimer.

Bollet, Alfred J. 1992. "Politics and Pellagra: The Epidemic of Pellagra in the U.S. in the Early Twentieth Century." *Yale Journal of Biology and Medicine* 65 (3): 211–21.

Boyd, Richard. 1979. "Metaphor and Theory Change: What is 'Metaphor' a Metaphor for?" In *Metaphor and Thought*, pp. 356–408. A. Ortony, ed. Cambridge, UK: Cambridge University Press.

Boyden, Edward S., F. Zhang, E. Bamberg, G. Nagel, K. Deisseroth. 2005. "Millisecond-timescale, Genetically Targeted Optical Control of Neural Activity." *Nature Neuroscience* 8:1263–68.

Boyle, Robert. [1772] 1968. "A Disquisition about the Final Causes of Natural Things." In *The Works of the Honourable Robert Boyle.* pp. 393–444. London: Thomas Birch.

Brenner, S., F. Jacob, and M. Meselson. 1961. "An Unstable Intermediate Carrying Information From Genes to Ribosomes for Protein Synthesis." *Nature* 190:576–81.

Bridewell, Will. 2004. "Science as an Anomaly-Driven Enterprise: A Computational Approach to Generating Acceptable Theory Revisions in the Face of Anomalous Data." PhD Dissertation. Department of Computer Science, University of Pittsburgh, Pittsburgh, PA.

Bridgman, Percy Williams. 1927. *The Logic of Modern Physics.* New York: Macmillan.

Brigandt, Ingo and Alan Love. 2008. "Reductionism in Biology." In *The Stanford Encyclopedia of Philosophy*. Edward N. Zalta, ed. http://plato.stanford.edu/entries/reduction-biology.

Browne, Janet. 1995. *Charles Darwin: Voyaging*. Princeton University Press.

———. 2002. *Charles Darwin: The Power of Place*. New York: Knopf.

Burchfield, Joe D. 1990. *Lord Kelvin and the Age of the Earth*. Chicago: University of Chicago Press.

Burian, Richard M. 1996. "Some Epistemological Reflections on Polistes as a Model Organism." In *Natural History and Evolution of an Animal Society: The Paper Wasp Case*. pp. 318–37. S. Turillazzi and M. J. West-Eberhard, eds. Oxford: Oxford University Press.

Bylebyl, Jerome J. 1973. "The Growth of Harvey's *De motu cordis*." *Bulletin of the History of Medicine* 47:427–70.

———. 1977. "*De motu cordis*: Written in Two Stages: Response." *Bulletin of the History of Medicine* 51:140–50.

———. 1982. "Boyle and Harvey on the Valves in the Veins." *Bulletin of the History of Medicine* 5:351–57.

Calcott, Brett. 2009. "Lineage Explanations: Explaining How Biological Mechanisms Change." *The British Journal for the Philosophy of Science* 60 (1): 51–78.

Castle, W. E. 1906. "Yellow Mice and Gametic Purity." *Science*, n.s., 24:275–81.

Castle, W. E. and C. C. Little. 1910. "On a Modified Mendelian Ratio among Yellow Mice." *Science* 32:868–70.

Chang, Hasok. 2007. *Inventing Temperature: Measurement and Scientific Progress*. New York: Oxford University Press.

———. 2009. "Operationalism." *The Stanford Encyclopedia of Philosophy*. Edward N. Zalta, ed. http://plato.stanford.edu/operationalism.

Chapuis, N., M. Durup, and C. Thinus-Blanc. 1987. "The Role of Exploratory Experience in a Shortcut Task by Golden Hamsters (*Mesocricetus auratus*)." *Animal Learning and Behavior* 15:174–78.

Chargaff, Erwin. 1950. "Chemical Specificity of Nucleic Acids and Mechanism of Their Enzymatic Degradation. " *Experientia* 6:201–09.

Chemero A. and M. Silberstein. 2008. "After the Philosophy of Mind: Replacing Scholasticism with Science." *Philosophy of Science* 75:1–27.

Chmiel, James F. and Pamela B. Davis. 2003. "Inflammatory Responses in the Cystic Fibrosis Lung." In *The Cystic Fibrosis Transmembrane Conductance Regulator*, pp. 160–80. Kevin L. Kirk and David C. Dawson, eds. New York: Kluwer.

Chow, Alan Y., Vincent Y. Chow, Kirk H. Packo, John S. Pollack, Gholam A. Peyman, Ronald Schuchard. 2004. "The Artificial Silicon Retina Microchip for the Treatment of Vision Loss From Retinitis Pigmentosa." *Archives of Ophthalmology* 122:460–69.

Churchland, Patricia. 1989. *Neurophilosophy: Toward a Unified Science of the Mind-Brain*. Cambridge, MA: MIT Press.

Cranefield, Paul F. 1957. "The Organic Physics of 1847 and the Biophysics of Today." *Journal of the History of Medicine and Allied Sciences* 12 (4): 407–23.

Craver, Carl F. 2001. "Role Functions, Mechanisms, and Hierarchy." *Philosophy of Science* 68:53–74.

———. 2002. "Interlevel Experiments, Multilevel Mechanisms in the Neuroscience of Memory." *Philosophy of Science* 69 (Proceedings): S83–S97.

———. 2003. "The Making of a Memory Mechanism." *Journal of the History of Biology* 36:153–95.

———. 2004. "Dissociable Realization and Kind Splitting." *Philosophy of Science* 71:960–71.

———. 2005. "Beyond Reduction: Mechanisms, Multifield Integration, and the Unity of Neuroscience." In *Studies in History and Philosophy of Biological and Biomedical Sciences.* Carl F. Craver and Lindley Darden, eds. Special Issue: "Mechanisms in Biology" 36:373–97.

———. 2006. "When Mechanistic Models Explain." *Synthese* 153:355–76.

———. 2007. *Explaining the Brain: Mechanisms and the Mosaic Unity of Neuroscience.* New York: Oxford University Press.

———. 2008a. "Physical Law and Mechanistic Explanation in the Hodgkin and Huxley Model of the Action Potential." *Philosophy of Science* 75 (5): 1022–33.

———. 2008b. "Axelrod, Julius." In *New Dictionary of Scientific Biography.* Noretta Koertge, ed. 1:118–23. Detroit, MI: Charles Scribner's Sons/Thomson Gale.

———. 2009. "Mechanisms and Natural Kinds." *Philosophical Psychology* 22:575–94.

———. 2010. "Prosthetic Models." *Philosophy of Science* 77:840–51.

———. Forthcoming. "Functions and Mechanisms: A Perspectivalist View." In *Functions: Selections and Mechanisms.* Philippe Huneman, ed. Dordrecht: Springer.

Craver, Carl F. and Lindley Darden. 2001. "Discovering Mechanisms in Neurobiology: The Case of Spatial Memory." In *Theory and Method in the Neurosciences,* pp. 112–37. Peter Machamer, R. Grush, and P. McLaughlin, eds. Pittsburgh: University of Pittsburgh Press. Reprinted in Darden 2006, ch. 2.

———. 2005. "Introduction: Mechanisms Then and Now." In *Studies in History and Philosophy of Biological and Biomedical Science.* Carl F. Craver and Lindley Darden, eds. Special Issue, "Mechanisms in Biology" 36:233–44.

Crick, Francis H. C. 1958. "On Protein Synthesis." *Symposium of the Society of Experimental Biology* 12:138–63.

———. 1959. "The Present Position of the Coding Problem." *Structure and Function of Genetic Elements: Brookhaven Symposia in Biology* 12:35–39.

———. 1967. "The Origin of the Genetic Code." *Nature* 213:119.

———. 1988. *What Mad Pursuit: A Personal View of Scientific Discovery.* New York: Basic Books.

———. 1996. "The Impact of Linus Pauling on Molecular Biology." In *The Pauling Symposium: A Discourse on the Art of Biography,* pp. 3–18. Ramesh S. Krishnamurthy, ed. Corvallis, Oregon: Oregon State University Libraries Special Collections.

Cuénot, Lucien. 1905. "Les Races Pures et Leurs Combinaisons Chez Les Souris." *Archives de Zoologie Expérimentale et Générale* 4 Serie. t.111:123–32.

Cummins, Robert. 1975. "Functional Analysis." *Journal of Philosophy* 72:741–64.

Darden, Lindley. 1976. "Reasoning in Scientific Change: Charles Darwin, Hugo de Vries, and the Discovery of Segregation." *Studies in History and Philosophy of Science* 7:127–69.

———. 1986. "Relations among Fields in the Evolutionary Synthesis." In *Integrating Scientific Disciplines,* pp. 113–23. W. Bechtel, ed. Dordrecht: Nijhoff. Reprinted in Darden 2006, ch. 7.

———. 1987. "Viewing the History of Science as Compiled Hindsight." *AI Magazine* 8 (2): 33–41.

————. 1991. *Theory Change in Science: Strategies from Mendelian Genetics.* New York: Oxford University Press.

————. 1995. "Exemplars, Abstractions, and Anomalies: Representations and Theory Change in Mendelian and Molecular Genetics." In *Concepts, Theories, and Rationality in the Biological Sciences,* pp. 137–58. James G. Lennox and Gereon Wolters, eds. Pittsburgh: University of Pittsburgh Press. Reprinted in Darden 2006, ch. 10.

————. 2002. "Strategies for Discovering Mechanisms: Schema Instantiation, Modular Subassembly, Forward/Backward Chaining." *Philosophy of Science* 69, Proceedings, S354–S365.

————. 2005. "Relations Among Fields: Mendelian, Cytological, and Molecular Mechanisms." In *Studies in History and Philosophy of Biological and Biomedical Sciences.* Carl F. Craver and Lindley Darden, eds. Special Issue: "Mechanisms in Biology" 36:349–71. Reprinted in Darden 2006, ch.4.

————. 2006. *Reasoning in Biological Discoveries: Mechanisms, Interfield Relations, and Anomaly Resolution.* New York: Cambridge University Press.

————. Forthcoming. "Mechanisms versus Causes in Biology and Medicine." In *Mechanism and Causality in Biology and Economics.* Hsiang-Ke Chao, Szu-Ting Chen, and Roberta L. Millstein, eds. The Netherlands: Springer.

Darden, Lindley and Joseph A. Cain. 1989. "Selection Type Theories." *Philosophy of Science* 56:106–29. Reprinted in Darden 2006, ch. 8.

Darden, Lindley and Carl F. Craver. 2002. "Strategies in the Interfield Discovery of the Mechanism of Protein Synthesis." *Studies in History and Philosophy of Biological and Biomedical Sciences* 33:1–28. Corrected and reprinted in Darden 2006, ch. 3.

Darden, Lindley and Nancy Maull. 1977. "Interfield Theories." *Philosophy of Science* 44:43–64. Reprinted in Darden 2006, ch. 5.

Darwin, Charles. 1859. *On the Origin of Species by Means of Natural Selection, or the Preservation of Favoured Races in the Struggle for Life.* London: John Murray.

————. [1859] 2009. *The Annotated Origin: A Facsimile of the First Edition of the Origin of Species.* Annotated by James T. Costa. Cambridge, MA: Harvard University Press.

————. 1868. *The Variation of Plants and Animals Under Domestication.* 2 vols. New York: Orange Judd and Co.

Darwin, Francis, ed. [1892] 1958. *The Autobiography of Charles Darwin and Selected Letters.* New York: Dover.

Datteri, Edoardo. 2009. "Simulation Experiments in Bionics: A Regulative Methodological Perspective." *Biology and Philosophy* 24 (3): 301–24.

Datteri, Edoardo and Guglielmo Tamburrini. 2007. "Biorobotic Experiments for the Discovery of Biological Mechanisms." *Philosophy of Science* 74:409–30.

Dawkins, Richard. 1976. *The Selfish Gene.* New York: Oxford University Press.

————. 2006. *The Selfish Gene.* 3rd ed. New York: Oxford University Press.

Deisseroth Karl. 2010. "Controlling The Brain with Light." *Scientific American* 303:48–55. doi:10.1038/scientificamerican1110–48.

————. 2011. "Optogenetics." *Nature Methods* 8:26–29. doi:10.1038/nmeth.f.324. Published online 2010.

Del Castillo J. and B. Katz. 1957. "The Study of Curare Action with an Electrical Micro-Method." *Proceedings of the Royal Society, London B Biological Sciences* 146:339–56. doi: 10.1098/rspb.1957.0015.

DesAutels, Lane. 2011. "Against Regular and Irregular Characterizations of Mechanisms." *Philosophy of Science* 78:914–25.

Des Chene, Dennis. 2001. *Spirits & Clocks: Machine & Organism in Descartes*. Ithaca, NY: Cornell University Press.

Descartes, René. [1649] 2000. *Passions of the Soul*. Excerpts reprinted in *Philosophical Essays and Correspondence (Descartes)*. Roger Ariew, ed. Indianapolis, IN: Hackett.

———. 1998. *The World [Le Monde] and Other Writings*. Stephen Gaukroger, ed. New York: Cambridge University Press.

Dietrich, Michael R. and Robert A. Skipper, Jr. 2007. "Manipulating Underdetermination in Scientific Controversy: The Case of the Molecular Clock." *Perspectives on Science* 15 (3): 295–326.

Dijksterhuis, E. J. 1961. *The Mechanization of the World Picture*. New York: Oxford University Press.

Dobzhansky, Theodosius. 1937. *Genetics and the Origin of Species*. New York: Columbia University Press.

Dretske, Fred I. 1994. "If You Can't Make One, You Don't Know How It Works." *Midwest Studies in Philosophy* 19 (1): 468–82.

Driesch, Hans. 1929. *Man and the Universe*. London: Allen and Unwin.

Drumm, Mitchell L. 2001. "The Race to Find the Cystic Fibrosis Gene: A Trainee's Inside View." In *Cystic Fibrosis in the 20th Century: People, Events, and Progress* pp. 79–92. Carl F. Doershuk, ed. Cleveland, OH: AM Publishing.

Dubos, René J. 1976. *The Professor, The Institute, and DNA: Oswald T. Avery, His Life and Scientific Achievements*. New York: Rockefeller University Press.

Dunbar, Kevin. 1995. "How Scientists Really Reason: Scientific Reasoning in Real-World Laboratories." In *The Nature of Insight*, pp. 365–95. R.J. Sternberg and J. E. Davidson, eds. Cambridge, MA: MIT Press.

Dunn, Leslie C. 1965. *A Short History of Genetics*. New York: McGraw-Hill.

Ebbinghaus, H. 1885. *Über das Gedächtnis. Untersuchungen zur experimentellen Psychologie*. Leipzig: Duncker & Humblot. English edition 1913. *Memory. A Contribution to Experimental Psychology*. New York: Teachers College, Columbia University. Reprinted Bristol: Thoemmes Press, 1999.

Eberhardt, F., P. O. Hoyer, and R. Scheines. 2010. "Combining Experiments to Discover Linear Cyclic Models with Latent Variables." *Journal of Machine Learning, Workshop and Conference Proceedings*. AISTATS 2010 9:185–92.

Elliott, Kevin. 2004. "Error as Means to Discovery." *Philosophy of Science* 71:174–97.

———. 2006. "A Novel Account of Scientific Anomaly: Help for the Dispute over Low-Dose Biochemical Effects." *Philosophy of Science* 73:790–802.

Frankel, Henry R. 2012. *The Continental Drift Controversy*. 4 vols. Cambridge, UK: Cambridge University Press.

Forterre, Yoël, Jan M. Skotheim, Jacques Dumais, and L. Mahadevan. 2005. "How the Venus Flytrap Snaps." *Nature* 433:421–25.

French, Roger. 1994. *William Harvey's Natural Philosophy*. Cambridge, UK: Cambridge University Press.

Futuyma, Douglas J. 2005. *Evolution*. Sunderland, MA: Sinauer Associates.

Gabbey, Allan. 1985. "The Mechanical Philosophy and its Problems: Mechanical Explanations, Impenetrability, and Perpetual Motion." In *Change and Progress in Modern Science*, pp. 9–84. Joseph C. Pitt, ed. Dordrecht: Reidel.

———. 1990. "The Case of Mechanics: One Revolution or Many?" In *Reappraisals of the Scientific Revolution*, pp. 493–528. David C. Lindberg and Robert S. Westman, eds. New York: Cambridge University Press.

Galen. 1997. *Galen: Selected Works* Translated with notes by P. N. Singer. New York: Oxford University Press.

Galileo, Galilei. [1610] 1989. *Sidereus Nuncius or The Sidereal Messenger.* Translated by Albert van Helden. Chicago: University of Chicago Press.

Gall, Franz J. and J. C. Spurzheim [1810–1819] 1835. *On the Functions of the Brain and Each of Its Parts.* Translated by W. Lewis, Jr. Boston, MA: Marsh, Capen, and Lyon.

Galton, Francis. 1871. "Experiments in Pangenesis, by Breeding From Rabbits of a Pure Variety, into Whose Circulation Blood Taken From Other Varieties had Previously been Largely Transfused." *Proceedings of the Royal Society (London)* 19:393–410.

Garber, Daniel. 1992. *Descartes' Metaphysical Physics.* Chicago: University of Chicago Press.

Gefter, Amanda. 2010. "Newton's Apple: The Real Story." *The New Scientist.* Culture Lab Blog. January 18, 2010. http://www.newscientist.com/blogs/culturelab/2010/01/newtons-apple-the-real-story.html.

Geison, Gerald. 1969. "Darwin and Heredity: The Evolution of His Hypothesis of Pangenesis." *Journal of the History of Medicine and Allied Sciences* 24:375–411.

Gentner, Dedre. 1983. "Structure Mapping–A Theoretical Framework for Analogy." *Cognitive Science* 7:155–70.

———. 1997. "Structure Mapping in Analogy and Similarity." *American Psychologist* 52:45–56.

Gentner, Dedre, Sarah Brem, Ron Ferguson, Phillip Wolff, Arthur B. Markman, and Ken Forbus. 1997. "Analogy and Creativity in the Works of Johannes Kepler." In *Creative Thought: An Investigation of Conceptual Structures and Processes*, pp. 403–59. T. B. Ward and J. Vaid, eds. Washington, D.C.: American Psychological Association.

Giere, Ronald. 1997. *Understanding Scientific Reasoning.* 4th ed. Fort Worth, Texas: Harcourt Brace College Publishers.

Glennan, Stuart S. 1992. *Mechanisms, Models, and Causation.* PhD dissertation, University of Chicago. Chicago, IL.

———. 1996. "Mechanisms and The Nature of Causation." *Erkenntnis* 44:49–71.

———. 2002. "Rethinking Mechanistic Explanation." *Philosophy of Science* 69 (Proceedings): S342–S353.

———. 2009. "Productivity, Relevance and Natural Selection." *Biology and Philosophy* 24 (3): 325–39.

Goel, Ashok and B. Chandrasekaran. 1989. "Functional Representation of Designs and Redesign Problem Solving." In *Proceedings of the Eleventh International Joint Conference on Artificial Intelligence*, pp. 1388–94. Detroit, MI.

Goldberger, Joseph. "The War on Pellagra." National Institutes of Health. Office of NIH History. Accessed November 24, 2012. http://history.nih.gov/exhibits/goldberger.

Goldberger, J. and G. A. Wheeler. 1915. "Experimental Pellagra in the Human Subject Brought about by a Restricted Diet." *Public Health Reports* 30:3336.

Gros, Francois, Howard Hiatt, Walter Gilbert, Chuck G. Kurland, R.W. Risebrough, and James D. Watson. 1961. "Unstable Ribonucleic Acid Revealed By Pulse Labeling of E. coli." *Nature* 190:581–85.

Haken, H., J. A. S. Kelso, and H. Bunz. 1985. "A Theoretical Model of Phase Transitions in Human Hand Movements." *Biological Cybernetics* 51:347–56.

Hanson, Norwood Russell. [1961] 1970. "Is There a Logic of Scientific Discovery?" In *Current Issues in the Philosophy of Science*. H. Feigl and G. Maxwell, eds. New York: Holt, Rinehart and Winston. Reprinted in *Readings in the Philosophy of Science*. pp. 620–33. B. Brody, ed. Englewood Cliffs, NJ: Prentice Hall.

Hanson, Norwood Russell. 1963. *The Concept of the Positron: A Philosophical Analysis*. Cambridge, UK: Cambridge University Press.

Hargrave, Paul A. 2001. "Rhodopsin." *Encyclopedia of the Life Sciences*. John Wiley & Sons. http://www.els.net. doi:10.1038/npg.els.

Harvey, William. 1963. *The Circulation of the Blood, and Other Writings*. Translated by Kenneth C. Franklin. New York: Everyman's Library, Dutton.

Hedström, Peter. 2005. *Dissecting the Social: On the Principles of Analytical Sociology*. Cambridge, UK: Cambridge University Press.

Hedström, Peter and Petri Ylikoski. 2010. "Causal Mechanisms in the Social Sciences." *Annual Review of Sociology* 36:49–67.

Hedström, Peter and Richard Swedberg, eds. 1998. *Social Mechanisms: An Analytical Approach to Social Theory*. Cambridge, UK: Cambridge University Press.

Hesse, Mary. 1966. *Models and Analogies in Science*. Notre Dame, IN: University of Notre Dame Press.

Hoagland, Mahlon B. 1955. "An Enzymic Mechanism for Amino Acid Activation in Animal Tissues." *Biochimica et Biophysica Acta* 16:288–89.

———. 1959. "Nucleic Acids and Proteins." *Scientific American* 201 (6): 55–61.

———. 1990. *Toward the Habit of Truth*. New York: Norton.

———. 1996. "Biochemistry or Molecular Biology? The Discovery of 'Soluble RNA.'" *Trends in Biological Sciences Letters (TIBS)* 21:77–80.

Hoagland, Mahlon B., Paul Zamecnik, and Mary L. Stephenson. 1959. "A Hypothesis Concerning the Roles of Particulate and Soluble Ribonucleic Acids in Protein Synthesis." In *A Symposium on Molecular Biology*, pp. 105–14. R. E. Zirkle, ed. Chicago: University of Chicago Press.

Hobbes, Thomas. [1655] 1981. *De Corpore*. Chapters 1–6. In Part I of *De Corpore*. Translated by A. P. Martinich. New York: Abaris Books.

Hodgkin, A. L. 1992. *Chance and Design: Reminiscences of Science in Peace and War*. Cambridge, UK: Cambridge University Press.

Hodgkin, A. L. and A. F. Huxley. 1952. "A Quantitative Description of Membrane Current and Its Application to Conduction and Excitation in Nerve." *Journal of Physiology* 117:500–44.

Holyoak, Keith J. and Paul Thagard. 1995. *Mental Leaps: Analogy in Creative Thought*. Cambridge, MA: MIT Press.

Hooker, C. A. 1981a. "Towards a General Theory of Reduction. Part I: Historical and Scientific Setting." In *Dialogue* 20 (1): 38–59.

———. 1981b. "Towards a General Theory of Reduction. Part II: Identity in Reduction." In *Dialogue* 20 (2): 201–36.

———. 1981c. "Towards a General Theory of Reduction. Part III: Cross-Categorical Reduction." In *Dialogue* 20 (3): 496–529.

Howick, Jeremy. 2011. "Exposing the Vanities—and a Qualified Defense—of Mechanistic Reasoning in Health Care Decision Making." *Philosophy of Science* 78:926–40.

Howick, Jeremy, Paul Glasziou, Jeffery K. Aronson. 2010. "Evidence-based Mechanistic Reasoning." *Journal of the Royal Society of Medicine* 103:433–41. doi: 10.1258/jrsm.2010.100146.

Hull, David. 1974. *Philosophy of Biological Science*. Englewood Cliffs, NJ: Prentice-Hall.

Humayun, Mark S., Eugene de Juan, Jr., Gislin Dagnelie, Robert J. Greenberg, Roy H. Propst, D. Howard Phillips. 1996. "Visual Perception Elicited by Electrical Stimulation of Retina in Blind Humans." *Archives of Ophthalmology* 114 (1): 40–46.

Huxley, A. F. 1963. "The Quantitative Analysis of Excitation and Conduction in Nerve." Nobel lecture. http://nobelprize.org/medicine/laureates/1963/huxley-lecture.html.

Huxley, Julian. 1942. *Evolution: The Modern Synthesis*. London: G. Allen & Unwin.

Jacob, Francois. 1988. *The Statue Within: An Autobiography*. Translated by Franklin Philip. New York: Basic Books.

Josephson, John R. and Susan G. Josephson, eds. 1994. *Abductive Inference: Computation, Philosophy, Technology*. New York: Cambridge University Press.

Judson, Horace F. 1996. *The Eighth Day of Creation: The Makers of the Revolution in Biology*. Expanded Edition. Cold Spring Harbor, NY: Cold Spring Harbor Laboratory Press.

Kandel, E. R., J. H. Schwartz, and T. M. Jessell. 1990. *Principles of Neural Science*. New York: McGraw Hill.

Kaplan, David and Carl F. Craver. 2011. "The Explanatory Force of Dynamical and Mathematical Models in Neuroscience: A Mechanistic Perspective," *Philosophy of Science* 78 (4): 601–27.

Karp, Peter. 1989. Hypothesis Formation and Qualitative Reasoning in Molecular Biology. PhD dissertation. Stanford University. Stanford, CA.

———. 1990. "Hypothesis Formation as Design." In *Computational Models of Scientific Discovery and Theory Formation*, pp. 275–317. J. Shrager and P. Langley, eds. San Mateo, CA: Morgan Kaufmann.

Kauffman, Stuart A. 1971. "Articulation of Parts Explanation in Biology and the Rational Search for Them." In *PSA 1970*, Boston Studies in the Philosophy of Science, vol. 8, pp. 257–72. Roger C. Buck and Robert S. Cohen, eds. Dordrecht: Reidel, Reprinted In *Topics in the Philosophy of Biology*. 1976, pp. 245–63. Marjorie Grene and Everett Mendelsohn, eds. Dordrecht: Reidel.

Kelvin, William Thomson. 1862a. "On the Age of the Sun's Heat." *Macmillan's Magazine* 5:288–393. Reprinted in *Popular Lectures and Addresses*, vol. 2, pp. 349–68.

———. 1862b. "On the Secular Cooling of the Earth." *Transactions of the Royal Society of Edinburgh* 23:157–69. Reprinted in *Mathematical and Physical Papers*, vol. 3, pp. 295–311.

———. 1865. "The Doctrine of Uniformity in Geology Briefly Refuted." *Proceedings of the Royal Society of Edinburgh* 5:512. Reprinted in *Popular Lectures and Addresses*, vol. 2, pp. 6–7.

———. 1882–1911. *Mathematical and Physical Papers*. 6 vols. Cambridge, UK: Cambridge University Press.

———. 1891–1894. *Popular Lectures and Addresses*. 3 vols. London: Macmillan and Co.

Kemeny, J. and P. Oppenheim. 1956. "On Reduction." *Philosophical Studies* 7:6–17.

Kendler, Kenneth S., P. Zachar, and Carl F. Craver. 2011. "What Kinds of Things are Psychiatric Disorders?" *Psychological Medicine* 21 (6): 1143–50.

Keyes, Martha. 1999. "The Prion Challenge to the 'Central Dogma' of Molecular Biology, 1965–1991, Part I: Prelude to Prions." *Studies in History and Philosophy of Biological and Biomedical Sciences* 30:1–19.

Kirk, Kevin L. and David C. Dawson, eds. 2003. *The Cystic Fibrosis Transmembrane Conductance Regulator*. New York: Kluwer.

Kitcher, Philip. 1984. "1953 and All That: A Tale of Two Sciences." *The Philosophical Review* 93:335–73.

Koonin, E. V. and A. S. Novozhilov. 2009. "Origin and Evolution of the Genetic Code: The Universal Enigma." *IUBMB Life*. Journal of the International Union of Biochemistry and Molecular Biology: Wiley Interscience. 61 (2): 99–111.

Kuhn, Thomas. 1962. *The Structure of Scientific Revolutions*. Chicago: University of Chicago Press.

Lakatos, Imre. 1970. "Falsification and the Methodology of Scientific Research Programmes." In *Criticism and the Growth of Knowledge*, pp. 91–195. I. Lakatos and Alan Musgrave, eds. Cambridge, UK: Cambridge University Press.

Lakatos, Imre. 1976. *Proofs and Refutations: The Logic of Mathematical Discovery*. J. Worrall and E. Zahar, eds. Cambridge, UK: Cambridge University Press.

Lamarck, Jean B. [1809] 1963. *Zoological Philosophy*. Translated by Huge Elliot. Chicago: University of Chicago Press.

Lebedev, M. A. and M. A. L. Nicolelis. 2006. "Brain Machine Interfaces: Past, Present, and Future." *Trends in Neurosciences* 29:536–46.

Lee, M. R. 2005. "Curare: The South American Arrow Poison." *Journal of the Royal College of Physicians, Edinburgh* 35:83–92.

Lennox, James G. 2006a. "William Harvey's Experiments and Conceptual Innovation." *Medicina & Storia* 12:5–27.

———. 2006b. "The Comparative Study of Animal Development: William Harvey's Aristotelianism." In *The Problem of Animal Generation in Modern Philosophy*, pp. 21–46. Justin Smith, ed. Cambridge, UK: Cambridge University Press.

Lennox, James G. and Bradley E. Wilson. 1994. "Natural Selection and the Struggle for Existence" *Studies in History and Philosophy of Science* 25:65–80.

Lewontin, Richard C. 1970. "The Units of Selection." *Annual Review of Ecology and Systematics* 1:1–18.

Lindee, Susan. In Preparation. "A Disease About to Disappear: Cystic Fibrosis and Gene Therapy in the early 1990s." Preliminary findings presented at the History of Science Society Meeting. Montreal 2010.

Lindee, Susan and Rebecca Mueller. 2011. "Is Cystic Fibrosis Genetic Medicine's Canary?" *Perspectives in Biology and Medicine* 54 (3): 316–31.

Lloyd, Elizabeth. 1987. "Confirmation of Ecological and Evolutionary Models." *Biology and Philosophy* 2:277–93.

———. 2007. "Units of Selection." In *Cambridge Companion to Philosophy of Biology*, pp. 44–65. David L. Hull and Michael Ruse, eds. New York: Cambridge University Press.

Machamer, Peter. 2004. "Activities and Causation: The Metaphysics and Epistemology of Mechanisms." *International Studies in the Philosophy of Science* 18:27–39.

Machamer, Peter, Lindley Darden, and Carl F. Craver. 2000. "Thinking About Mechanisms." *Philosophy of Science* 67:1–25. (Referred to as MDC.) Reprinted in Darden 2006, ch. 1.

Malthus, Thomas. [1798] 1999. *An Essay on the Principle of Population*. Geoffrey Gilbert, ed. New York: Oxford World Classics, Oxford University Press.

Marieb, Elaine N. 2004. *Human Anatomy & Physiology*. 6th ed. San Francisco: Benjamin Cummings.

Mayo, Deborah G. 1996. *Error and the Growth of Experimental Knowledge*. Chicago: University of Chicago Press.

Mayr, Ernst. 1942. *Systematics and the Origin of Species*. New York: Columbia University Press.

McCarty, Maclyn. 1985. *The Transforming Principle: Discovering that Genes are Made of DNA*. New York: Norton.

McKusick, Victor A. 1969. "On Lumpers and Splitters, or The Nosology of Genetic Disease." *Perspectives in Biology and Medicine* 12 (2): 298–312.

Mendel, Gregor. [1865] 1966. "Experiments on Plant Hybrids." In *The Origin of Genetics, A Mendel Source Book*, pp. 1–48. Curt Stern and Eva Sherwood, eds. San Francisco: W.H. Freeman.

Moghaddam-Taaheri, Sara. 2011. "Understanding Pathology in the Context of Physiological Mechanisms: The Practicality of a Broken-Normal View." *Biology and Philosophy* 26:603–11. doi:10.1007/s10539–011–9258–2.

Monod, Jacques. [1965] 1977. "From Enzymatic Adaptation to Allosteric Transitions." Reprinted in *Nobel Lectures in Molecular Biology 1933–1975*, pp. 259–82. New York: Elsevier.

Morange, Michel. 2006. "The Ambiguous Place of Structure Biology in the Historiography of Molecular Biology." In *History and Epistemology of Molecular Biology and Beyond: Problems and Perspectives*, pp. 179–86. Hans-Jörg Rheinberger and Soraya de Chadarevian, eds. Berlin: Max Planck Institute for the History of Science.

Morgan, Thomas H. 1905. "The Assumed Purity of the Germ Cells in Mendelian Results." *Science* 22:877–79.

Morgan, Thomas H., A. H. Sturtevant, H. J. Muller, and C. B. Bridges. 1915. *The Mechanism of Mendelian Heredity*. New York: Henry Holt and Company.

Mowry, Bryan. 1985. "From Galen's Theory to William Harvey's Theory: A Case Study in the Rationality of Scientific Theory Change." *Studies in the History and Philosophy of Science* 16 (1): 49–82.

Nagel, Ernest. 1961. *The Structure of Science*. New York: Harcourt, Brace and World.

Neander, Karen. 1991. "Functions as Selected Effects: The Conceptual Analyst's Defense." *Philosophy of Science* 58:168–84.

Newsome III, George L. 2003. "The Debate Between Current Versions of Covariation and Mechanism Approaches to Causal Inference." *Philosophical Psychology* 16 (1): 87–107.

Newton-Smith, William H. 1981. *The Rationality of Science*. Boston: Routledge & Kegan Paul.

Nervi, Mauro. 2010. "Mechanisms, Malfunctions and Explanation in Medicine." *Biology and Philosophy* 25:215–28. doi:10.1007/s10539–009–9190-x.

Olby, Robert. 2009. *Francis Crick: Hunter of Life's Secrets*. Cold Spring Harbor, NY: Cold Spring Harbor Press.

Olton, David S. and Robert J. Samuelson. 1976. "Remembrance of Places Passed: Spatial Memory in Rats." *Journal of Experimental Psychology: Animal Behavior Processes* 2:97–116.

Ospovat, Dov. 1981. *The Development of Darwin's Theory*. Cambridge, UK: Cambridge University Press.

Parascandola, Mark. 1995. "Philosophy in the Laboratory: The Debate Over Evidence for E.J. Steele's Lamarckian Hypothesis." *Studies in History and Philosophy of Science* 26:469–92.

Pardee, Arthur B. 1979. "The PaJaMa Experiment." In *Origins of Molecular Biology: A Tribute to Jacque Monod*, pp. 109–16. Andre Lwoff and Agnes Ullmann, eds. New York: Academic Press.

Pardee, Arthur B., Francois Jacob, and Jacques Monod. 1959. "The Genetic Control and Cytoplasmic Expression of 'Inducibility' in the Synthesis of β-galatosidase." *Journal of Molecular Biology* 1:165–78.

Paul, Diane B. 1988. "The Selection of the 'Survival of the Fittest.'" *Journal of the History of Biology* 21:411–24.

Pearl, Judea. 2000. *Causality: Models, Reasoning, and Inference.* New York: Cambridge University Press.

Petroski, Henry. 1992. *To Engineer is Human: The Role of Failure in Successful Design.* New York: Random House, Vintage Books.

Plato. 1961. *Meno, Phaedo, and Phaedrus.* In *The Collected Dialogues of Plato.* Edith Hamilton and Huntington Cairns, eds. Princeton University Press.

Ramsey, Jeffry L. 2008. "Mechanisms and Their Explanatory Challenges in Organic Chemistry." *Philosophy of Science* 75 (5): 970–82.

Rheinberger, Hans-Jörg. 1997. *Towards a History of Epistemic Things: Synthesizing Proteins in the Test Tube.* Palo Alto, CA: Stanford University Press.

Roediger, H. L. 2008. "Relativity of Remembering: Why the Laws of Memory Vanished." S. Fiske, ed. In *Annual Review of Psychology* 59:225–54.

Rudge, David W. 1999. "Taking the Peppered Moth with a Grain of Salt." *Biology & Philosophy* 14:9–37.

———. 2001. "Kettlewell from an Error Statistician's Point of View." *Perspectives on Science* 9:59–77.

Ruse, Michael. 1973. "The Value of Analogical Models in Science." *Dialogue* 12:246–53.

———. 1979. *The Darwinian Revolution: Science Red in Tooth and Claw.* Chicago: University of Chicago Press.

Russ, Rosemary, Rachel E. Scherr, David Hammer, and Jamie Mikeska. 2008. "Recognizing Mechanistic Reasoning in Student Scientific Inquiry: A Framework for Discourse Analysis Developed From Philosophy of Science." Special Issue: Science Studies and Science Education. *Science Education* 92 (3): 499–524. doi:10.1002/sce.20264.

Russ, Rosemary, Janet E. Coffey, David Hammer, and Paul Hutchinson. 2009. "Making Classroom Assessment More Accountable to Scientific Reasoning: A Case for Attending to Mechanistic Thinking." *Science Education* 93 (5): 875–91. doi:10.1002/sce.20320.

Salmon, Wesley C. 1984. *Scientific Explanation and the Causal Structure of the World.* Princeton University Press.

Schacter, D. L. and E. Tulving, eds. 1994. *Memory Systems 1994.* Cambridge, MA: MIT Press.

Schaffner, Kenneth. 1974. "Logic of Discovery and Justification in Regulatory Genetics." *Studies in History and Philosophy of Science* 4:349–85.

———, ed. 1985. *Logic of Discovery and Diagnosis in Medicine.* Berkeley: University of California Press.

————. 1993. *Discovery and Explanation in Biology and Medicine*. Chicago: University of Chicago Press.

————. 2008. "Theories, Models, and Equations in Biology: The Heuristic Search for Emergent Simplifications in Neurobiology." *Philosophy of Science* 75:1008–21.

Seeley, Robin Hadlock. 1986. "Intense Natural Selection Caused a Rapid Morphological Transition in a Living Marine Snail." *Proceedings of the National Academy of Sciences (USA)* 83:6897–901.

Sexton, Renard J. 2007. Unpublished. "Public Policy Implications of Environmental Mechanisms." Presented at the Meeting of the International Society for History, Philosophy, and Social Studies of Biology, Exeter, UK.

Shapin, Steven. 1996. *The Scientific Revolution*. Chicago: University of Chicago Press.

Simon, Herbert A. 1996. *The Sciences of the Artificial*. 3rd ed. Cambridge, MA: MIT Press.

Skipper, Jr., Robert A. and Roberta L. Millstein. 2005. "Thinking about Evolutionary Mechanisms: Natural Selection." In *Studies in History and Philosophy of Biological and Biomedical Sciences*. Carl F. Craver and Lindley Darden, eds. Special Issue: "Mechanisms in Biology" 36:327–47.

Spencer, Herbert. 1864. *Principles of Biology*. 2 vols. London: Williams and Norgate.

Spirtes, Peter, Clark Glymour, and Richard Scheines. 2001. *Causation, Prediction, and Search*. 2nd ed. Cambridge, MA: MIT Press.

Steel, Daniel P. 2008. *Across the Boundaries: Extrapolation in Biology and Social Science*. New York: Oxford University Press.

Steele, Edward J., Robyn A. Lindley, and Robert B. Blanden. 1998. *Lamarck's Signature*. Reading, MA: Perseus Books.

Strain, Daniel. 2011. "Evolution's Wedges: Finding the Genes that Drive One Species into Two." *Science News* 180, Issue 1, July 2, p. 18.

Sullivan, Jacqueline A. 2009. "The Multiplicity of Experimental Protocols: A Challenge to Reductionist and Non-Reductionist Models of the Unity of Neuroscience." *Synthese* 167:511–39.

Tabery, James G. 2004. "Synthesizing Activities and Interactions in the Concept of a Mechanism." *Philosophy of Science* 71:1–15.

Temin, Howard M and Satoshi Mizutani. 1970. "RNA-dependent DNA Polymerase in Virions of Rous Sarcoma Virus." *Nature* 226:1211–13.

Thagard, Paul. 1998. "Explaining Disease: Causes, Correlations, and Mechanisms." *Minds and Machines* 8:61–78.

————. 1999. *How Scientists Explain Disease*. Princeton, NJ: Princeton University Press.

————. 2006. *Hot Thought: Mechanisms and Applications of Emotional Cognition*. Cambridge, MA: MIT Press, Bradford Books.

————. 2012. *The Cognitive Science of Science: Explanation, Discovery, and Conceptual Change*. Cambridge, MA: MIT Press.

Tolman, Edward C. 1948. "Cognitive Maps in Rats and Men." *Psychological Review* 55:189–208.

Tolman, Edward C. and C. Honzik. 1930. "Introduction and Removal of Reward and Maze Performance in Rats." *University of California Publications in Psychology* 5:257–75.

Trumpler, Maria. 1997. "Converging Images: Techniques of Intervention and Forms of Representation of Sodium-Channel Proteins in Nerve Cell Membranes." *Journal of the History of Biology* 30:55–89.

Tufte, Edward. 1998. *Visual Explanations: Images and Quantities, Evidence and Narrative.* Cheshire, CT: Graphics Press.

Valenstein, Elliot S. 2005. *The War of the Soups and the Sparks: The Discovery of Neurotransmitters and the Dispute Over How Nerves Communicate.* New York: Columbia University Press.

Wang, Xiaodong et al. 2006. "Hsp90 Cochaperone aha1 Downregulation Rescues Misfolding of CFTR in Cystic Fibrosis." *Cell* 127:803–15.

Waterton, Charles. 1825. *Wanderings in South America, the North-West of the United States, and the Antilles, in the years 1812, 1816, 1820, and 1824: With Original Instructions for the Perfect Preservation of Birds, &c. for Cabinets of Natural History.* London: J. Mawman.

Watson, James D. [1962] 1977. "The Involvement of RNA in the Synthesis of Proteins." In *Nobel Lectures in Molecular Biology 1933–1975*, pp. 179–203. New York: Elsevier.

———. 1963. "Involvement of RNA in the Synthesis of Proteins." *Science* 140:17–25.

———. 1968. *The Double Helix.* New York: New American Library.

Watson, James D., Tania A. Baker, Stephen P. Bell, and Alexander Gann. 2007. *Molecular Biology of the Gene.* 6th ed. San Francisco, CA: Benjamin Cummings.

Watson, James D. and Francis H. C. Crick. 1953a. "Molecular Structure of Nucleic Acids: A Structure for Deoxyribose Nucleic Acid." *Nature* 171:737–38.

———.1953b. "Genetical Implications of the Structure of Deoxyribonucleic Acid." *Nature* 171:964–67.

Weber, Marcel. 2008. "Causes without Mechanisms: Experimental Regularities, Physical Laws, and Neuroscientific Explanation." *Philosophy of Science* 75:995–1007.

Westfall, Richard S. 1971. *The Construction of Modern Science: Mechanisms and Mechanics.* New York: John Wiley and Sons.

Whitteridge, Gweneth. 1971. *William Harvey and the Circulation of the Blood.* London: Macdonald.

———. 1977. "*De Motu Cordis*: Written in Two Stages." *Bulletin of the History of Medicine* 51:130–40.

Wilson, Margaret. 1999. *Ideas and Mechanism: Essays on Early Modern Philosophy.* Princeton University Press.

Wilson, Robert A. and Frank C. Keil. 1998. "The Shadows and Shallows of Explanation." *Minds and Machines* 8:137–59.

Wimsatt, William. 1972. "Complexity and Organization." In *PSA 1972, Proceedings of the Philosophy of Science Association*, pp. 67–86. Kenneth F. Schaffner and Robert S. Cohen, eds. Dordrecht: Reidel.

———. 1980. "Reductionist Research Strategies and Their Biases in the Units of Selection Controversy." In *Scientific Discovery: Case Studies*, pp. 213–59. T. Nickles, ed. Dordrecht: Reidel.

———. 1987. "False Models as Means to Truer Theories." In *Neutral Models in Biology*, pp. 23–55. Matthew Nitecki and Antoni Hoffman, eds. New York: Oxford University Press.

———. 1997. "Aggregativity: Reductive Heuristics for Finding Emergence." *Philosophy of Science* 64. Proceedings: S372–S384.

———. 2007. *Re-Engineering Philosophy for Limited Beings: Piecewise Approximations to Reality.* Cambridge, MA: Harvard University Press.

Woodward, James. 2003. *Making Things Happen: A Theory of Causal Explanation.* New York: Oxford University Press.

———. 2011. "Data and Phenomena: A Restatement and Defense." *Synthese* 182 (1): 165–79.

Wouters, Arno. 2005. "The Functional Perspective of Organismal Biology." In *Current Themes in Theoretical Biology*, pp. 33–69. T. A. C. Reydon and L. Hemerik, eds. The Netherlands: Springer.

Wright, Larry. 1973. "Functions." *Philosophical Review* 82:139–68.

———. 1976. *Teleological Explanations*. Berkeley: University of California Press.

Wylie, Alison. 2002. *Thinking from Things: Essays in the Philosophy of Archaeology*. Berkeley, CA: University of California Press.

Zamecnik, Paul C. 1960. "Historical and Current Aspects of the Problem of Protein Synthesis." *The Harvey Lectures 1958–1959*. Series 54, pp. 256–81. New York: Academic Press.

INDEX

Abrahamsen, Adele, 27, 183
abstraction: degrees of, 33, 37; vs.
    generality, 36; in mathematical models,
    42; process of forming, 32, 51; in visual
    representations, 38
action potentials, 42, 43–50, 110–11
activity-enabling properties, 78–79
activity signatures, 78
adaptation: instructive-type schema for,
    76–77; selection-type schema for, 74
Allchin, Douglas, 159
Allen, Colin, 63
Allen, Garland E., 14, 82
analogies, 71–72, 74–75, 81
Anand, Bal Krishan, 55–56, 63
Andersen, Holly, 28
anomalies. *See* revision of schemas
Archimedes, 64
Aristotle, 3, 5, 97, 115, 117
Avery, Oswald, 132–33, 142–43
Axelrod, Julius, 129, 134–37, 140, 142

Bacon, Francis, 3, 14, 141, 142, 195
bacteria: *E. coli*, 70, 138–39, 156;
    Pneumococcus, 132–33; virulence of,
    132–33
Baker, Jason, 185
Baltimore, David, 82
Barker, Matthew J., 18, 185
Barros, D. Benjamin, 28, 185
Beatty, John, 27
Bechtel, William: on heuristics, 26; on
    interfield integration, 183; mecha-
    nistic philosophy and, 27, 28; on
    reconstituting phenomena, 62; on
    strategies for discovering mecha-
    nisms, 81
Bender, Edward A., 142
Benfey, O. T., 81

Bernard, Claude: curare poisoning and,
    1–2, 13; on homeostatic regulation,
    72–73; inhibitory experiments of, 127;
    mechanistic biology and, 4
bestiaries, 5
Bickle, John, 184
biology: aims of science and, 6; homun-
    cular explanations in, 88–89; interfield
    collaboration in, 21–22; lack of laws
    in, 27; levels of mechanisms and,
    21–22; maker's knowledge in, 93;
    mathematical and graphical models
    in, 42, 50; mechanistic view in, 3–7,
    12–13, 26–27, 97, 141, 197; Mendel,
    Gregor, and Mendelian genetics and,
    173–77, 184; model organisms and
    systems in, 141; nonmechanistic
    explanations in, 87; pragmatic power
    of mechanisms in, 12–13, 186, 195;
    preparing experimental systems in,
    137–41; structures of entities and,
    106; types of mechanisms and, 81. *See
    also* interfield integration
Birmingham, A. T., 13
Black, John, 13
Bliss, T. V. P., 168–69, 184
blood circulation: between arteries and
    veins, 115–16; constraints on mecha-
    nisms and, 200–201; Galenic view of,
    98–100, 103–8, 116–17; historical
    understandings of, 98–102; motion of
    the heart and, 112; multilevel mecha-
    nisms and, 23–24; parts of body and,
    102–4; pulse and, 109–10, 112; timing
    of activities in, 112–13; valves and,
    105–8, 114; William Harvey and, 3, 97
Boas, Marie, 13
Bogen, James, 28, 51, 63
Bollet, Alfred J., 142

boxes, black, gray, and glass: by-what-activity experiments and, 129–31; by-what-entity experiments and, 131–33; completeness of schemas and, 31–33, 59, 89–90, 144; filler terms and, 31; goal of life sciences and, 173–75; interlevel experiments and, 125; mechanism discovery as piecemeal affair and, 8; productive continuity and, 19, 65, 83; protein synthesis and, 154, 158; role anomalies and, 153; sketches vs. schemas and, 199–200

boxology, 90–91

Boyd, Richard, 81

Boyden, Edward S., 195

Boyle, Robert, 97, 105–6

brain studies: activation experiments in, 128–29; fPET and fMRI studies and, 128–29; functional imaging and, 124–25; homuncular explanations and, 88; inhibiting conditions and, 57–58; inhibitory experiments in, 127, 129; lateral hypothalamus and, 55–56; localization of phenomena and, 103; norepinephrine and, 134–37; spatial memory and, 52–54; stimulation experiments in, 128. See also neurotransmitters

Brenner, S., 160

Bridewell, Will, 160

Bridgman, Percy Williams, 63

Brigandt, Ingo, 183

Brobeck, John R., brain studies of, 55–56, 63

Browne, Janet, 82

Bunz, H., 95

Burchfield, Joe D., 95

Burian, Richard, on falsifying anomalies, 159

Bylebyl, Jerome J., 117

Byron, Jason, 185

Caenorhabditis elegans, 161, 162

Cain, Joseph A., 51, 82, 185

Calcott, Brett, 28

Castle, William, 150–51, 152, 160

cause: Aristotelian, 5, 7; Bacon and role of in science, 142, 195; experimenta-

tion and, 119–25, 129, 131, 134, 140, 201; as filler term, 31; formal methods for finding, 11, 14; new mechanistic analysis of, 26–27; as represented by variables, 41

CFTR gene, 190–92, 195

chaining, forward and backward, 77–80, 111

chainology, 91–92, 111, 115

Chandrasekaran, B., 160

Chang, Hasok, 63

Chapuis, N., 63

characterization of phenomena: constraints on mechanisms and, 56, 97–98, 199–200; data and experimentation and, 54–56; discovery of mechanisms and, 8, 11, 52–56; fictional phenomena and, 60–61; how-possibly and how-plausibly schemas and, 200; inhibiting conditions and, 57–59; lumping and splitting and, 61–62; mischaracterization and, 60; modulating conditions and, 58; nonstandard conditions and, 58–59; precipitating conditions and, 56–57; by-products and, 59–60; recharacterization and, 60–62; taxonomic errors in, 61–62

Chargaff, Erwin, 133

Chemero, A., 95

Chesapeake Bay, dead zones in, 186, 187–89

Churchland, Patricia, 62

clonal selection, 74, 82

cocaine, 136–37

Collingridge, G. L., 184

Colombo, Matteo Realdo, 98, 117

construction of schemas: from aha to strategy in, 64–65; analogies and, 71–72, 74–75, 80; forward and backward chaining and, 77–80; guidance from phenomenon and store, 67–69; how-possibly and how-plausibly schemas and, 200; localization and, 69–71, 80; modular subassembly and, 74–77, 80; phenomenal models vs. mechanistic schemas and, 87–89; product shapes discovery and, 65–66; schema types and, 71–74; selection schemas and, 73–74; in stages of mechanism discovery, 8, 11; types of mechanisms and, 67–69

Cranefield, Paul F., 13

Craver, Carl: on constraints in reasoning about mechanisms, 118; on contemporary views of mechanisms, 14; on discovery of messenger RNA, 160; on filler terms, 51; on Hodgkin-Huxley model, 51; on interfield integration, 183–84; on levels of mechanisms, 27, 29; on lumping and splitting, 63; on making and knowing, 96; on multilevel experiments in neuroscience, 142; on PaJaMa experiments, 143; on phenomenal vs. explanatory models, 95; on protein synthesis; on role-function, 63

Crick, Francis: on discovery of messenger RNA, 160; DNA research of, 32, 51, 78–79, 175–76; Linus Pauling and, 184; protein synthesis and, 35, 82, 164, 165; on ribosomal anomaly, 160

Cuénot, Lucien, 148–50, 160

Cummins, Robert, 29, 63

curare, 1–2

curiosity, xii

cystic fibrosis, 186–87, 189–93, 195

Dale, Henry, 134

Darden, Lindley: on abstraction for natural selection, 51; on anomalies, 159, 160; on constraints in reasoning about mechanisms, 118; 26–28, 183, 185, on contemporary views of mechanisms, 14; on cystic fibrosis, 195; on discovery of messenger RNA, 160; on evolutionary theories, 96; on interfield integration 183–84, 185; on multilevel experiments in neuroscience, 142; on PaJaMa experiments, 143; on protein synthesis, 82; on reasoning strategies, 81; on selection theories, 82; on T. H. Morgan's work, 82; on virtues of theories, 95; on working entities in mechanisms, 29

Darwin, Charles: analogy to domestic breeding and, 71–72; causes of variation and, 89; definition of fitness and, 185; evolutionary synthesis and, 177–78, 181–82; history of evolutionary theory and, 184; mechanistic biology and, 4; natural selection and 71–72, 184; pangenesis and, 95–96; principle of divergence and, 181–82; questions asked by, 73; as rule breaker, 85; survival of the fittest and, 185; Thomas Malthus and, 72, 82

Darwin, Francis, 82

data, characterization of phenomena and, 54–56

Datteri, Edoardo, 96

Dawkins, Richard, 178, 184

Dawson, David C., 195

dead zones, 186, 187–89

decomposition, 21, 24, 26, 81, 183

Deisseroth, Karl, 195

Del Castillo, J., 13

DesAutels, Lane, 28

Descartes, René, 3–5, 13, 97

Des Chene, Dennis, 13

description of mechanisms, 89

determinism, regularity of mechanisms and, 20

dietary deficiency, 120–25

Dietrich, Michael R., 95

Dijksterhuis, E. J., 13

discovery. See search for mechanisms

DNA: bacterial transformation and, 132–34; discovery of structure of, 175–76; enzyme induction and, 138–41; forward chaining and, 78–79; fowl tumors and, 133; genetic code and, 36; genetic medicine and, 190–92; historical study of, 70, 133; Lamarckian adaptation and, 77; modularity and, 75–76, 77; molecular biology and, 173–76; mutations in, 178, 190–91; protein synthesis and, 31–33, 79, 154–58, 164–67; replication of, 17–20, 37; viruses and, 147

Dobzhansky, Theodosius, 178–79, 182, 184

Dretske, Fred I., 96

Driesch, Hans, 88–89, 95
Drosophila melanogaster, 176, 178
Drumm, Mitchell L., 195
du Bois-Reymond, Emil, 4, 13
Dubos, René, 143
Dunbar, Kevin, 81
Dunn, Leslie C., 96

Earth, age of, 85
Ebbinghaus, Herman, 59, 63
Eberhardt, F., 142
ecology, 128, 186, 187–89, 195
electrical circuits, 46–50
Elliott, Kevin, 159
empirical constraints on schemas:
    abilities of components and, 106–9;
    activities of components and, 109–11;
    blood circulation and, 100–102; global
    organization and, 115–17; localization
    of phenomena and, 102–4; productivity
    flow and, 114; reasoning and, 9–10;
    role of components and, 114–15; struc-
    tures of entities and, 104–6; timing of
    activities and, 111–13
enzyme induction, 138–41
equilibrium: equilibrium potential and,
    45–48; nonlinear mechanisms and, 19
eutrophication, 187–89, 195
evaluation of schemas: abilities of com-
    ponents and, 106–9; breaking rules
    in, 85; "Build it" test and, 92–94, 96;
    coherence and, 83–84, 85; complete-
    ness and, 86–94; conservatism and, 84;
    consideration of alternative schemas
    and, 146; correctness and, 9, 83,
    94–95, 97–98; elegance and, 84; em-
    pirical adequacy and, 84, 85; explana-
    tory value of, 84, 85; fertility and, 84;
    generality of, 84, 86; "How does that
    work" test and, 90–91; knowledge-how
    vs. knowledge-that and, 93; prediction
    and, 84, 85; simplicity and, 84, 85–86;
    in stages of mechanism discovery, 8–9,
    11; superficiality and, 86–89; testability
    and, 83, 84–85; unification and, 84, 86;
    virtues of good theories and, 83–86;

"What if that worked differently" test
    and, 91–92
evolution: causes of variation and, 89;
    evolutionary synthesis and, 177–78;
    from gene's perspective, 178–79;
    history of theory of, 184; interfield
    integration and, 172; Lamarckian,
    76–77; modularity in, 75, 77; modular
    systems of the body and, 61; reduc-
    tion in, 178–79; reuse of old modules
    within mechanisms and, 37; selection
    schemas and, 73–74. See also adapta-
    tion; Darwin, Charles; Dobzhansky,
    Theodosius; natural selection
explanation: as description
    of mechanism, 31, 87, 89;
    deductive-nomological (DN) model
    of, 27
experiments: activation, 128; back-
    ground knowledge and, 125;
    bottom-up vs. top-down, 125–26; by-
    what-activity, 129–31, 133; by-what-
    entity, 131–33; for causal relevance,
    119–25, 129, 140, 201; for compo-
    nential relevance, 125–29; detection
    techniques and, 124–25; excitatory vs.
    inhibitory interventions and, 126–28,
    140; experimental systems and, 122;
    interference, 126–27; interlevel,
    125–29; intervention techniques and,
    122–24; as interviews with nature,
    141–42; multiple interventions in,
    133–37, 140–41; nonstandard condi-
    tions in, 59; phenomena and data and,
    54–56; preparation of systems for,
    137–41; relevance of techniques in,
    55–56; series of, 133–37; stimulation,
    127–28; types of, 119, 142, 201

Fabricius ab Acquapendente, 98, 105,
    107, 114
failures of schemas, 86–90
fertilization, visual representation of,
    39–41
fields of science: of biology, 161;
    distinguishing features of, 161; as not

coextensive with size levels, 21. *See also* interfield integration.

filler terms, 31

finger-wagging model, 87–88

Foterre, Yoël, 142

Frankel, Hank, 28

Franklin, Kenneth C., 117

French, Roger, 117

Freud, Sigmund, 60

functions: proper, 24, 53–54, 63, 70; represented by gray boxes, 31, 89, 129, 131–33, 153–54, 158, 199, 200; roles of in mechanisms, 23–24, 63, 153, 155–56, 190–91

Futuyma, Douglas J., 184–85

Gabbey, Allan, 13

Galen of Pergamon, 3, 98–100, 103–10, 114, 116–17

Gall, Franz Joseph, 62

Galton, Francis, 96

Garber, Daniel, 13

Gardner-Medwin, A. R., 168–69

Geison, Gerald, 96

genetically modified organisms, 93, 186

genetic code: Francis Crick and, 51, 168; as frozen, 36–37; *Pax6* gene and, 75

genetic inheritance: adaptive traits and, 76; anomalous schemas in, 147; bacterial transformation and, 132; chromosomes and, 147; evolutionary synthesis and, 177–78, 181–82; forward chaining and, 78–79; gene's perspective on, 178–79; historical understandings of, 69–70, 71, 95–96, 179; incomplete schemas and, 89; isolating mechanisms and, 182; vs. learned responses, 76; Mendelian segregation and, 149–51; selfish genes and, 178; superficial models and, 88–89. *See also* evolution

genetic medicine, 186–87, 189–94, 195

genetic studies: bacterial transformation and, 132–33; enzyme induction and, 138–41; inhibitory experiments in, 127; interfield integration in, 172–73;

monster anomalies and, 148–51; stimulation experiments in, 128

Gentner, Dedre, 81

Giere, Ronald, 95

Glennan, Stuart, 26, 27–28

Goel, Ashok, 160

Goldberger, Joseph, 120–25, 142

Gros, Francois, 160

Haken, H., 95

Hanson, Norwood Russell, 51, 81

Hargrave, Paul A., 95

Harvey, William, 3, 9–10, 97–110, 112–18, 168, 200–201

Hedström, Peter, 28–29

heredity. *See* genetic inheritance; genetic studies

Hertting, Georg, 135–36

Hesse, Mary, 81

heuristics, 26, 160, 198. *See also* construction of schemas; evaluation of schemas; revision of schemas

HIV-AIDS, 75–76, 93

Hoagland, Mahlon, 82, 165–66, 178–79, 184

Hobbes, Thomas, 97

Hodgkin, Alan L., 42, 43–50, 51, 110–11, 167–68

Holyoak, Keith J., 81

homeostatic regulation, 66, 72–73

homuncular explanations, 88–89

Honzick, C., 63

Hooker, C. A., 62

"How does that work" test, 90–91

Howick, Jeremy, 28

Hubbard, L. Ron, 60

Hull, David, 184

human anatomy: blood circulation and, 102–4; diagrams of, 99, 100, 101; historical understandings of, 98–102; localization of phenomena and, 103–4; motion of the heart and, 109–11; structures of entities and, 104–6

Huxley, Andrew F., 42, 43–50, 51, 110–11, 167–68

Huxley, Julian, 184

Immunology: clonal selection and, 74, 82; neutrophils and cystic fibrosis, 192
inheritance, genetic. See genetic inheritance
interfield integration: continuous intertemporal, 177–82; interlevel, 163, 167–72, 177–82; intertemporal, 163–64, 180, 182–83; protein synthesis and, 164–67; reduction and, 183; roundworms and, 161–62; scaffolding mechanisms and, 201; sequential intertemporal, 172–77; simple, single level, 163–67

Jacob, Francois, 138–41, 143, 156, 160
jellyfish, osmoregulation in, 70
Josephson, John R., 81
Josephson, Susan G., 81
Judson, Horace F., 160

Kandel, E. R., 95, 184
Kaplan, David, 51, 95
Karp, Peter, 160
Katz, B., 13
Kauffman, Stuart, 26
Keil, Frank C., 96
Kekulé, August, 64, 81
Kelso, Scott, 87–88, 96
Kelvin (Lord; Sir William Thomson), 74, 85, 95, 111
Kemeney, J., 183
Kendler, Kenneth S., 63
Kettlewell, H. B. D., 168, 185
Kirk, Kevin L., 195
Kitcher, Philip, 184
knowledge-how vs. knowledge-that, 93, 186
Koonin, E. V., 51
Krebs cycle, 18–19
Kuhn, Thomas, 159
Kuhnian crisis anomalies, 159. See also revision of schemas

lactose metabolism, 70
Lakatos, Imre, 159–60
Lamarck, Jean-Baptiste, 76–77, 179
learning: inhibition of, 57–58; multilevel,

multifield perspective on, 184; synaptic plasticity and, 168–72
Lee, M. R., 13
Lennox, James G., 117, 185
level of mechanism vs. size 21–22. See also interlevel integration
Lewontin, Richard, 185
light transduction, 92–93
Lindee, Susan, 195
Little, C. C., 160
Lloyd, Elizabeth, 95, 185
Loewi, Otto, 95, 110, 130–31, 133–34, 138, 185
Lømo, Terje, 168–69
Long-Term Potentiation, 168–72
Love, Alan, 183
LTP. See Long-Term Potentiation

Machamer, Peter, 14, 26–27, 28
machines: vs. mechanisms, 15–16; molecular, 51
MacLeod, C. M., 142
malnutrition, 120–25
Malthus, Thomas, 72, 81
Maull, Nancy, 183
Mayo, Deborah G., 159
Mayr, Ernst, 182, 185
McCarty, Maclyn, 142–43
McGuire, Ted, 14
McKusick, Victor, 63
mechanisms: characterization of, 15, 199; components and features of, 16–25; contexts of, 24–25; decompositional view and, 26; difference-making conditions in, 120; entities and activities and, 16–18, 78; functions in, 23–24; global organization of, 115–17; hierarchy of, 167–69; levels of, 21–23; linear vs. nonlinear, 18–19; machines vs., 15–16, 51; mechanistic joints and, 62; multilevel, multifield perspective on, 25, 162–63, 167–72, 179–82, 184; natural selection as, 28, 179; nearly decomposable systems and, 167; organization of, 20; productive continuity and, 19; productivity of, 113–14; regularity of, 19–20; roles of components of, 114–

lies and, 147, 151–52, 157, 158–59; modular redesign and, 157–59; monster anomalies and, 146–47, 148–51, 159–60; organizational anomalies and, 153; protein synthesis and, 154–56, 157–58; reasoning and, 160; role anomalies and, 153, 154, 155, 156; special case anomalies and, 147, 151–52, 157, 159; in stages of mechanism discovery, 9–10, 11–12; temporal anomalies and, 153, 154, 156; types of anomalies and, 145, 201

Rheinberger, Hans-Jörg, 82, 160

Richardson, Robert, 26, 62, 81

RNA: bacterial transformation and, 132–33; fowl tumors and, 133; genetic code and, 36; genetic medicine and, 190–91; Lamarckian adaptation and, 77; modularity and, 75–76, 77; protein synthesis and, 31–33, 79–80, 154–58, 167; as template, 79, 154–58, 165–66; viruses and, 147

Roediger, H. L., 63

roundworm. See *Caenorhabditis elegans*

Rudge, David, 185

Ruse, Michael, 82

Russ, Rosemary, 28

Salmon, Wesley, 27

Samuelson, Robert J., 63

Schacter, D. L., 63

Schaffner, Kenneth, 51, 62, 143, 160, 183, 184

schemas. See characterization of phenomena; construction of schemas; empirical constraints on schemas; evaluation of schemas; representation of mechanisms; revision of schemas; sketches

science: aha moments in, 64–65; characterization of mechanisms and, 199; community of scientists and, 146; definition of, 195, 201–202; history of, 74; mechanistic view of, 3; natural selection in, 185; philosophy of, 27–29, 183, 184; reasoning in, 160

science education: experiments in, 142; lack of interfield integration in, 183;

mechanistic reasoning in, 28; textbook schemas and, 32, 33, 167–68

search for mechanisms: aims of science and, 6–7; authors' goals and, 197–98; as conceptual and methodological, 2–3; discovery mechanisms and, 64–65; in fields other than biology, 198; formal vs. final causes and, 5; guidance from phenomenon and store of knowledge for, 67–69; historical origins of, 5–6; as integrative ideal in biology, xi; interference effects and, 59–60; limits on plausible solutions and, 68–69; localization of components and, 102–4; location of mechanisms and, 70–71; mechanical philosophy and, 4–5; mischaracterization of phenomena and, 60; occult vs. non-occult explanations and, 4–5; as pluralistic endeavor, 198; possibilities vs. live possibilities in, 68–69; process of discovery and, 65–66, 196; questions in, 65; reasons for, 26; as scaffold, 12; stages of, 7–10. See also mechanisms

Seeley, Robin Hadlock, 180, 185

selection, abstract schema for, 73–74. See also natural selection, clonal selection

Sexton, Renard J., 195

Shapin, Steven, 13

Silberstein, M., 95

Simon, Herbert, 61, 63, 167, 185

Singer, P. N., 117

sketches: anomalies and, 159; completeness and, 30–33, 91–92, 144; functional requirements and, 75; gaps in, 77, 89, 199–201; as hypotheses, 7; incorrect, 94; vs. schemas, 51, 56, 59–60, 67, 111. See also boxes, black, gray, and glass

Skipper, Robert A., Jr., 27, 28, 95

snail shells, natural selection and, 180–81

Snell's law, 87

Spencer, Herbert, 72, 82

Spirtes, Peter, 14

statistics, 145–46

Steel, Daniel P., 63